THE LOST GRIZZLIES

THE LOST GRIZZLIES

A Search for Survivors in the Wilderness of Colorado

> >

RICK BASS

A MARINER BOOK

HOUGHTON MIFFLIN COMPANY

BOSTON · NEW YORK

For information about permission to reproduce selections from
this book, write to Permissions, Houghton Mifflin Company,
215 Park Avenue South, New York, New York 10003.

Library of Congress Cataloging-in-Publication Data
Bass, Rick, date.
  The lost grizzlies : a search for survivors in the wilderness
of Colorado / Rick Bass.
    p.   cm.
  ISBN 0-395-71759-0   ISBN 0-395-85700-7 (pbk.)
  1. Grizzly bear — San Juan Mountains (Colo. and N.M.)
I. Title.
  QL737.C27B37   1995
  599.74'446 — dc20   95-21833   CIP

Printed in the United States of America

QUM 10 9 8 7 6 5 4 3 2 1

For information about this and other Houghton Mifflin
trade and reference books and multimedia products, visit
The Bookstore at Houghton Mifflin on the World Wide Web
at http://www.hmco.com/trade/.

The quotation on pages 6 and 7 is from *The Last Grizzly and
Other Southwestern Bear Stories,* edited by John A. Murray and
David E. Brown (Tucson: University of Arizona Press, 1988).
The quotation on pages 80 and 81 is from *Wildlife in Peril:
The Threatened and Endangered Mammals of Colorado,* by John
Murray (Boulder, Colo.: Robert Rinehart, Inc., 1987).

ACKNOWLEDGMENTS

I am grateful to many people for their help with this book, especially my friends in and out of the mountains and, once again, my editors, Larry Cooper and Camille Hykes. Deepest thanks to Russell Chatham for the elegant drawings, to my agent Bob Dattila, to Hilary Liftin for considerable production assistance, to John Murray for his exhaustive and inspired research, and to Melodie Wertelet for the book's design. Thanks are also due to Round River Conservation Studies. A portion of the royalties from sales of this book will go to help Round River, a nonprofit, tax-exempt ecological research and education organization. You can write to Round River at 4301 Emigration Canyon, Salt Lake City, Utah 84108.

A portion of this book appeared, in different form, in *Audubon* magazine.

FOR DOUG, TERRY, AND DENNIS

CONTENTS

PART I

›››››››››››››››››››

THE FALL

It was for this that he worshiped the bear. For man had fallen out of the secure world of instinct into a place of wonder.

— LOREN EISLEY

Cold autumn stars are shining down when I leave my wooded valley of Yaak, in northern Montana. I don't know if my old truck will make it all the way to southern Colorado, but I'm going to try. It's late September 1990. A friend, Doug Peacock, the grizzly bear expert, has told me he's been hearing reports and rumors that there are still grizzlies in Colorado, and he wants me to go into the mountains with him and look for them. The official government position is that they're extinct there.

I love my deep forest of Yaak, love the sanctuary it provides, but Doug, although a new friend, is already a dear one — a teacher and a guide — and so I head south. I don't really see how the grizzlies can be holding on down there. If nothing else, it'll be good to spend time with Doug in the woods.

Actually, I kind of believe they're down there. Hardly anyone else does — yet — and even Doug's not sure. But there is a place in our hearts for them, and so it is possible to believe they still exist, if only because that space of longing exists.

I drive with the windows down. It seems odd to be driving south to go look for grizzlies. Surely I'm bound for the wrong hemisphere, where water drains clockwise — or perhaps I'm traveling back in time, to my father's, or more likely my grandfather's, generation, back when there *were* grizzlies in Colorado.

The stars twinkle. I drive with my bare arm angled out the window. I pretend that it is 1940 and I am my father's father.

I drive all through the night, as if pushing back in time. I stop on a bridge above one lovely little creek and get out and walk down to the water's whispery, murmuring edge. The riffles throw back a disjointed reflection of the moon and stars. I crouch, take off my shirt, splash cold water on my face and shoulders, and scrub it into my hair. It's a cold night. I want to believe they're still down there. I want to find them.

*

By 1920, the last grizzlies had been exterminated in northern Colorado, but in southern Colorado they hung on through the thirties and forties. It was in 1950 that an old government trapper named Lloyd Andersen ("the Bear Man") killed a three-hundred-pound sow in this area, but her two cubs escaped. He killed another grizzly the following year, in the San Juan Mountains of southern Colorado, and the trapper Ernie Wilkerson killed a two-year-old male. This bear had dragged a twelve-foot spruce deadfall for five miles before Wilkerson could catch up with the bear and shoot it. Three years later, believing that all its grizzlies had been killed off, the state of Colorado formed the Rio Grande–San Juan Grizzly Bear Management Area to protect grizzlies, should any be seen.

Game wardens continued to report sightings. In 1954 another grizzly — a tiny one, no bigger than a good-size dog — was killed by a sheepherder. In 1955 "probable" signs of grizzlies were discovered. A game warden spotted a female and her cub in 1956. Lloyd Andersen found sets of tracks in 1957, and a skull was also found. That was thought for sure to be the end of them.

But the rumors kept coming, drifting down from a landscape that rises above ten thousand feet, the top of the world. Just rumors, though; no one was killing the grizzlies anymore — no more dead bears, which is always, sadly, the truest form of verification. (Otherwise, biologists will tell you, "Oh, you just saw the blond phase of a black bear.") People kept trying to say that the grizzlies were extinct in Colorado, but for the careful observer, the determined hiker, a few clues and signs could still be discovered.

The bears were still there, but it was as if they were becoming more secretive.

1962. Lloyd Andersen reported that a grizzly had killed twenty sheep, not far from the family ranch of a friend of Doug's and mine, the place where we're going to meet on this trip.

1964. Andersen, on horseback and with his dogs, chased a grizzly for twelve miles in the San Juans. The bear never treed but kept backing up against rocks and fighting the dogs. That grizzly was killed, proving the bears had been out there all along. Sightings reported from around this time take on special significance now, because grizzlies can live for twenty-five to thirty years in the wild.

1967. Andersen, camping again in southern Colorado, woke up and looked outside his tent — his horses were acting strange — and saw a sow grizzly with two yearlings working their way across a treeless slope. "They were finding pack rats' nests and playing with them and chasing each other. I suppose I watched 'em for about a half hour before they disappeared over the saddle about a half mile away," Andersen said. "No doubt about it. I've seen grizzlies before. As a government trapper, I trapped seven. Once you see a grizzly you won't forget it."

That same year, several other hunting parties in the San Juans also reported seeing a sow and two cubs, but that was it. Those were the last "good" sightings. In the early 1970s biologists set out dead horses, and in several instances bears moved those carcasses great distances. There have been numerous sightings since, but no confirmed sightings, with definitive photographs or dead grizzlies, and in 1975 John Torres of the Colorado Division of Wildlife stated, "Our records indicate that, for all practical purposes, the grizzly is now extirpated from Colorado."

*

On September 23, 1979, a bow hunter and guide named Ed Wiseman was near a high mountain lake in the San Juans with his client, Mike Niederee. The two were hunting elk. John Murray, in *The Last Grizzly,* writes:

At around five in the evening Niederee claims to have jumped a bear in its day bed. Despite being surprised at close distance, the bear chose to avoid a confrontation and fled the scene. Several hundred yards away it ran into Wiseman, and, according to their deposition, that is when the trouble began. The bear reportedly knocked Wiseman down and began to maul him. At first Wiseman tried to remain passive and appear dead, a strategy recommended in bear attacks. This, however, did not seem to work, and growing desperate, he grabbed a hunting arrow and began to stab the bear in the throat and neck. The bear, weakened by injuries, released Wiseman, lumbered off into the twilight, and died a short distance away.

Niederee, alerted by Wiseman's cries, arrived on the scene to find the outfitter a bloody mess. Wiseman had suffered several bites, a broken right leg, a mangled foot, and considerable blood loss in the attack. After applying what first aid he could, Niederee left to get the horses, which were tied off about one mile over the ridge. He returned with them, but Wiseman was in such bad shape that he could not be moved. Niederee made a large fire, gathered plenty of kindling, and, as the cold and darkness settled on the high mountains, left for Wiseman's hunting base camp. . . . Several hours later Niederee arrived at the base camp. The camp cook rode out to the trailhead for help, while the rest of the party, including Niederee's father, a surgeon, rode back . . . to Wiseman, reaching him about four in the morning. The rescue party rebuilt the fire and attended to his wounds. Wiseman had lost a lot of blood, but was not, in addition to his injuries, suffering from hypothermia. A helicopter evacuated him from the site later that morning. He eventually recovered from the attack. . . .

A later helicopter flight removed the grizzly bear skull and hide from the site, but left the bear's carcass behind. This was a substantial loss to science, as a close examination of the uterus for placental scars could have determined if the bear had ever borne cubs — evidence that other grizzlies might still reside in the area. A number of biologists who later examined the pelt agreed that the pigmentation and enlargement of the mammaries strongly suggested that the grizzly had nursed cubs in the past.

The Colorado Division of Wildlife conducted a two-year study of the South San Juans (1980–1982), and found only possible evidence of grizzlies in the area: a number of large digging sites, a partially collapsed grizzly bear den, and a possible sighting of a blond-phased adult female with two dark cubs. The last sighting was particularly intriguing because biologists found a large number of dig sites and a quantity of long blond bear hairs at that location. Black bears are generally thought to be unable to undertake large digs, as do grizzlies, because of their much shorter foreclaws. Because of the large size of the study area, the reclusiveness of grizzlies, the large number of black bears, and the ruggedness of the region, the results were inconclusive.

Inconclusive. I drift south, driving through the night. I wouldn't travel this far if there wasn't that spot, that place in my heart, that vacancy waiting to fill up with belief.

*

When Doug called, he gave me two days' notice. We're to meet at one P.M. on the last day of September at Betty Feazel's At Last Ranch, just outside Pagosa Springs, Colorado. Marty Ring, an artist who does work for, among other things, the Earth First! newsletters, will join us. We'll go into the woods and look for grizzlies ourselves. No more theorizing. We'll find out whether they're still there or not. Marty, on a previous trip, has seen what he thinks are claw marks, four deep scratches, up so high and cut so deep into the tree, he says, that surely they were made by a grizzly.

Betty Feazel is seventy-five years old, and from a Quaker family that has ranched in Wolf Creek Pass since the early 1900s. Betty, her daughter Lucy, and Lucy's husband, Bruce, an attorney, are in favor of protecting the grizzlies, if there are any that remain in Colorado. They say the day of range cattle is over, and that it costs more to raise a calf than what you can sell it for. The three of them are part of a committee, Save Our San Juans, which is working to help protect the wild quality of their home. It's the same old story of the West, the same one I can find in my own back yard — one I do not have to drive fifteen hundred miles to look for.

On the phone, Peacock had ranted to me about the biologists in charge of the San Juans. "They're okay guys," he said, "good guys, really, but Jesus, they just don't have any imagination." I could see him, a thousand miles away in his house in Tucson, slapping his half-bald head in exasperation and spinning in half-circles, getting tangled up in the phone cord, then swatting at it, cursing.

The next day, down in Colorado, when I reach Betty Feazel's ranch the aspens are as Doug had promised they'd be: in the full glory of autumn. I turn up the long gravel drive and see Peacock and Betty at the top. Peacock's walking around in shorts and hiking boots and a long flannel jacket. As usual, his wild thinning hair is askew. Dust still rises from the driveway; he's just arrived, too. He always reminds me of how good it feels to be alive. I get out of my truck and we shake hands, then hug.

"Wow," Peacock says, spreading his arms, "just look at this." Betty stands next to him, beaming, as if she'd planted these snow-crowned mountains herself. "Oh, shit," he says, and whirls around and stalks back to his Subaru. The tail end of the car nearly touches the ground under its load of bulging burlap sacks. "I forgot gifts. I was in Santa Fe last night. Jeez."

Empty beer cans, books on tape, canteens, and a pair of binoculars tumble out when he lifts the hatchback. He lunges, tries to catch each object as it falls.

"Aw, fuck it," he says, and crawls inside. The burlap sacks contain peppers, hundreds of big, red, spicy chili peppers. "Here," he says, wrestling one of the bags out. "To cook with. Where can I put this?"

Betty claps her hands, delighted. "In the basement," she says.

When I get within ten feet of Peacock's car, my eyes start to water and I sneeze.

"It was wonderful driving up here, breathing all that shit," he says, laboring beneath the bag. It looks like he's bought every pepper in Santa Fe. Just then Marty comes driving up in a dusty brown Volkswagen Rabbit. There is a big wolf-looking dog in the back seat of the car: his faithful, ancient Keetina.

We've come four thousand miles altogether, the three of us, and have arrived within ten minutes of one another. We go inside with Betty and Lucy and Bruce to reconnoiter — to look at photos and maps and listen to background information.

We'll only have a few days; I've got a grouse-hunting appointment with a friend back in Montana early the next week. This is mainly a trip to see and smell and feel the country, rather than to try to decide right away, yes or no, if there are any grizzlies left.

We sit like students at Betty's long dining room table. The farmhouse's windowpanes are wavy with age, bending the straw-colored sunlight that passes through them. Betty's energy is inspiring as she leans over the maps and points out areas where rumored sightings have occurred.

"Oh Jesus," Peacock says, looking at the topo maps. "Oh fuck, that's good country. Excuse me," he says to Betty, but she waves him off. "Oh fuck," he says again, "fuck yes."

The rumors are good. They're so similar to the rumors of the sixties and seventies that is seems certain a bear or bears are in there. A bear with a large hump was seen along Jo Jo Creek (most of the place names in the book have been changed to protect grizzly habitats). An outfitter saw what he believed were grizzly tracks — ten to twelve inches long, with big claws — along Blazo Creek in the summer of 1989. Sightings were reported along Wolf Creek Pass. The following year, Tony Povolitis, a senior scientist for the Humane Society, found a big track near Grizzly Creek and photographed it. We pass the photo around. It was taken in snow, and the track looks huge. We can tell where the claws landed. It's not conclusive, but it looks good.

Tom Beck is the biologist who headed the two-year Colorado Division of Wildlife study in the South San Juans in the wake of the Wiseman bear's "confirmation." Four trackers lived in tents from June to October, 1980 to 1982, and hunted for grizzly sign and set snares. Peacock bristles and huffs at this method; he's convinced that these bears won't be trapped and that all bears, especially grizzlies, know when they're being hunted.

Beck wrote, of that previous search, "Sometimes they'd trap a bear and really get excited because of its size, bleached tips on its brown fur, and the way it stood giving it a humped back, they'd swear they caught a grizzly." But all the trappers ever caught in their snares were black bears. Gary Gerhardt, a staff writer for the *Rocky Mountain News* who's been covering the rumors, wrote, "Beck said the division [of wildlife] assumes there aren't any grizzlies in Colorado, and it's going to take some strong documentation — such as a grizzly in a trap — for him to believe any are left in the San Juans."

One can imagine Peacock's terror at this philosophy. He's been trying to round up money to put a couple of fieldworkers in the San Juans, but he believes that they should engage in lower-impact methods to document grizzlies, such as photography or making casts of a track, but no trapping.

We could, of course, solve the problem quickly and simply. We could find grizzly sign or photograph a grizzly. I've got my camera. We could end all this political foolishness, all these abstractions. It might happen.

Now, in late September, the grizzlies are up high, getting ready to den. We aim to investigate the north-facing sides of the San Juans, and other places. We plan to do a lot of bushwhacking, crawling around in jungle and heavy windfall — which is not where bears typically hang out. But Peacock has a theory that any remaining San Juan grizzlies are atypical. They're smarter, and over the last forty years the mothers have raised their cubs to be solitary and to avoid humans. He's certain that these bears will by now be active primarily at night.

Before we begin to look for sign of them, we'll camp for two or three days at twelve thousand feet, so as to thin our blood and ready ourselves for the long hikes and heavy packs.

The next morning we say goodbye to Betty, Lucy, and Bruce and set off in two cars. We can feel their hopes riding with us like a net, a thing of a certain density, a specific weight. Of course the odds are long that in this brief trip we'll see a grizzly or find any sign of one, but there's always the chance. That empty place in our hearts.

He knows the habits of wild grizzlies better perhaps than anyone else in the country, and yet he abhors scientific meetings, academic conferences, and the like. He is an eloquent writer, and yet you absolutely cannot wring more than one or two sentences out of him concerning literature. He was a medic in Vietnam — a soldier, a warrior — and yet he dissolves into loving baby talk around children — around anyone under voting age.

For all of Peacock's complexity, however, he has basically two behavior patterns that I've observed. The first and most striking mode of behavior occurs when he's wired with an anxiety that leads to a mania reminiscent of the Bugs Bunny–Tasmanian Devil cartoons. When he's wound up that tight — twirling and blinking rapidly, owlishly, as if he can't believe how the world has turned against him — he'll invariably bolt, whether through a crowd of people or a heavy, tangled briar bush. Once spooked, he'll walk furiously away, his body seeking isobars of lower pressure — a calmer register of atmospheric conditions against his cells, is how I picture it. Peacock may be out of control, but it's nothing personal against the people who've spooked him. The second mode of behavior involves deep, unpretentious happiness, like the parody of a man in a beer commercial who takes a sip of fine brew, smacks his lips, and says "Ahhhhh!"

Thankfully, today opens with a grace note: we're bound for grizzly country. But first we must head to Pagosa for gas and more supplies.

Marty leads the way in his Rabbit. We pass an Exxon station, and I see Marty stick his hand out the window to give Exxon the finger. A minute later we pull into a Conoco station, pretending it's better. Peacock storms into the adjacent convenience store looking for condiments — Tabasco sauce and limes — and refuses to shop for the beer. "Just get anything," he says to Marty and me, staring up at the dull overhead lighting. His face wrinkles as he moves unsteadily on the linoleum between the narrow aisles. He knocks a can of chili off a shelf, shouts "Fuck!" and bolts, so Marty and I buy the beer and pay for the gas. We drive off with a feeling of having somehow escaped, as we race past trailer parks, amusement parks, and rectangular trout ponds where campers can fish for Rocky Mountain Rainbow Trout with Genuine Corn Pellets (if the campers are careful not to hook a southbound semi on their backcast).

With his "flee mechanism" still very much engaged, Peacock grumbles and curses, certain we've missed our turnoff. His mood is strangely contagious: we've all become tense. It's not until we find the correct dirt road and start up it, passing through fields of late summer columbine and lupine, through cool areas of shade that smell of ferns and moss and aspen, that he begins to recover, and we all relax.

"Look at this," Peacock says to me as the Subaru climbs. The sky is blue and the aspens glow an incandescent yellow. He smacks his forehead. "Isn't this unbelievable? Isn't this great?" The muscles in his face relax. His shoulders loosen.

We drink half a beer each while we drive. We've come from places at sea level and at three thousand feet, respectively. "I can't drink beer at altitude," Peacock explains, and he's right, everything's much too slow, much too weighted down already. Going up mountains, nearing the top of the world, the air engenders that slower-than-life feeling in which it can become difficult to move.

We ride up over the pass where, in the 1950s, men chased a grizzly with dogs and horses, but lost it. We move slowly around the crest of the mountains, and even our vehicles labor to breathe. There's a stillness above, so we know things inside us are being acted upon, both subtle and large, heart things. Our pace slows. We look for a campsite where we can stop to acclimate ourselves.

"Drink lots of water," Peacock advises. "*Gulp* water. You'll get sick as a dog if you don't." I've brought a big bottle of aspirin and I gulp those, too. The most troubling thing about altitude sickness for me is that once I've got it, it doesn't go away. I have to descend to make it better, and on this trip we won't be descending, we'll be climbing higher. The air is thin, as if in a science-fiction movie where the oxygen supply gets cut off. It's strange to think that a large animal can live and prosper at such a high elevation.

There's a shrill ringing in my ears that soon becomes a rough clatter. Am I drinking enough water? Alarmed, I open my canteen and guzzle, but as we round a switchback, I realize the clattering is coming from Marty's Volkswagen's rear wheel bearings. We pull over to consider it. The tops of the aspens wave. Billowing white clouds shoot past us. We're right at tree line, but we can't find any good camping spots where we can lie low and rest. The last thing we want to do is camp at the side of the road, where pilgrims' exhaust fumes can envelop us. We've got to camp up here somewhere to get those red blood cells expanding — to dilute ourselves, so to speak, with extra oxygen.

"We can fix that," Peacock says, pointing to Marty's car.

We're reassured. Neither Marty nor I knows enough about cars to change a set of windshield wipers, but Peacock dismisses the wheel-bearing problem with a total lack of concern. He won't even look at the car; he keeps gazing at the autumn ridges above us and extrapolating, I know, over those blue ridges, three or four days beyond them into the deeper wild.

"I'm sorry," Marty says. He's not worried about getting back to his job in Boulder, since he's on vacation, but he is mortified that he may slow down this mission. Doug shakes his head and says, "Fuck it. Fuck it. We'll fix that in two, three hours once we get back out." He looks up at the mountains and says, "Isn't this beautiful?"

We decide that the bearings are spun-on, heat-welded, and can't be hurt any further, and Marty allows as how the car's been making a funny noise since "back in Denver." Peacock winces, but then his face washes clear again, as peace-filled as a Buddhist's — *it's just a machine* — and we begin again our search for a copse, a hidden grove. The north side of the mountains will be more wooded and steeper, probably with running water. We continue driving, and before we realize it we're out of the high country and down in the river bottom, where we see corrals and buck-and-rail fences and campgrounds with trailers and a few tents. We turn around and head back up to where we've already been.

We take a side fork by a big yellow mined-out hill, which seems to glow with cyanide, and just around the bend from the hill we come upon the old gold mine's tailing pond, whose shores are also yellow and quicksand-murky, with a few dead-gray tree skeletons upright in the water. Miracle of miracles, trout rise from the polluted pond, making slow rolls at the late afternoon mayflies. Trout with three eyes, I think, trout with livers like small raisins and brains of solid gold. Perhaps their bones are lithified, composed also of gold, or maybe cyanide. Perhaps in the frying pan the fish would vaporize into cyanide gas and everyone in the house would fall down dead.

On the other side of the dark pond there's a handsome cabin set back in the trees. Somehow, someone lives here in the middle of the national forest. "How?" I ask, amazed. "They bought it," Peacock says matter-of-factly. In the middle of the trout pond there is a large yellow sign that says "Private Property — No Fishing — Stay In Your Car — This Means You!"

We stop in the middle of the road and stare at the sign as if it were written in a foreign language. Running-dog imperialist lackeys, I think, and try to smile. Peacock stays in the car, looking and looking, as if engaged in a stare-down with the sign, a contest to determine whether or not his time in the mountains will henceforth be ruined. Finally his breathing simmers down. He opens the door and deposits on the ground the few empty beer cans we've accumulated. Like coyotes leaving their spoor on a trap, we manage, without getting out of the car, to litter Cyanide Lake's private beach, and then we drive on, passing a logged-over campsite.

Ahead of us, Marty's car is roaring and groaning, squawking and shrieking. The day is ending. We've got to make camp soon.

At about eleven thousand feet, we cross a creek on the switchback and spot an opening in the bushes. Peacock drives off the road and through the opening, across a small, cow-grazed meadow and up into the lodgepole forest. Marty follows and parks on a flat spot where we can work on his car later. Steam and the smell of burning metal fill the air. I can't help but think that this will be the Rabbit's final resting place, that fifty years from now the Rabbit will resemble those rusted-out haul trucks one sometimes finds, with aspens growing through their middle, at the end of logging roads.

Purple thunderheads assemble to the north above the high rock reefs. A cool wet wind is in our faces. We gather wood and pitch our tents with privacy in mind, a good fifty yards from one another. Peacock spreads a ground tarp for his campsite — after more than two hundred nights spent in the woods last year, he's grown weary of confining tents. He studies the storm front sweeping our way and says, "It doesn't look like much rain. If it rains hard I'll get up and set up a tent. I hate fucking with tents."

Marty's dog, Keetina, bounds into the brush, wanting exercise after a long day of travel. While Marty and I build the fire, Peacock hauls food out of his backpack and tells us about his trip to Mexico early in the summer. He heard a jaguar and Mexican lobos in three different ranges, and he followed a good hard rumor of a Chihuahua grizzly. Then he tells us about the last time he slept in a tent, on Vancouver Island, and how he awoke in the night certain that he'd heard the sound of drums rising from Indian middens deep in the woods, and how the drumming was making him crazy, how it was telling him to take his knife and cut his tentmate's throat, and how he almost did it — the drumming was almost too strong and too terrible to overcome.

I note with relief that Marty is camped closer to Peacock than I am. I know that when I lie down to sleep, I will be listening for drumming in these woods, beyond the pounding of my own earth-anchored high-altitude heart.

Though he's a gourmet cook, Peacock saves his elegant cooking for when he's at home or visiting friends, with a real kitchen at his disposal. In the woods he mostly carries with him bags of tsamba, a simple dried-rice mix that's more suited to horses than people. I can't complain, though. All I ever carry is cans of hideous Vienna sausages, a habit I acquired when I worked in the oil fields. Marty, whom I like very much but who is, after all, from Boulder, and young, carries those space-age foil packets that pretend to be meals more nutritionally complete than what you'd ever eat at home: Octoroon Belgian Cheese-Batter Steak and Scalloped Mushrooms with Vichyssoise Apple Cinnamon Sauce.

But Peacock improvises. He pulls a fistful of hot peppers out of one of the burlap sacks, takes a chunk of the Swiss cheese Marty bought in the convenience store that afternoon, and demonstrates how to slice the peppers lengthwise, then peels shavings of cheese into the peppers and lays them next to the fire to roast. When the white cheese bubbles and the peppers sweat and glisten, we pick them up off the rocks, blow on them to cool them, hold them in cupped hands, and devour them. I've brought along a few small bottles of Wild Turkey, and we open one and sit there in the dark as a light rain begins to fall. We eat without utensils and pass the bottle, listen to the stream, put damp wood on the small fire, and tell stories.

Keetina races from the dark woods into the firelight looking like a large white wolf, and it's clear from her odor and brown-stained muzzle that she's found a dead something to roll in and feast upon.

"Fumarole!" Peacock shouts, trying to stay away from the dog's affectionate advances. Marty cries out more apologies, leads Keetina over to his tent downwind, ties her to an aspen tree, then returns to the fire. She whines, howls once, and lies down. We're all getting rained on.

It's a great meal, I muse, when you don't have to jack with knives and forks. I could live this way for a long time. I spend too much time at the typewriter.

Peacock tells us about the time he was without food, and utensils, in the Piedras Negras. He finally managed to catch some big lizards, kill them, and put them on the rocks by his fire to cook. "Didn't even have a knife," he says. "I had to wait for them to explode, then go gather them up. The meat was good." He pauses. "I was hungry then."

✳

I sleep all night with the sound of rain falling on the tent. The morning brings heavy fog over the mountains, and it's still raining lightly. My heart pounds from the thin air. I take a couple of aspirin for the altitude, a couple more for the Wild Turkey, and walk down to the stream to splash water on my face. The music of it reminds me of Wallace Stegner's book *The Sound of Mountain Water.* Doug is asleep on his back, inside his sleeping bag with a clear sheet of plastic pulled over him, his ski cap knocked sideways, looking like the worst kind of woods waif.

A hoarder by nature, I stashed dry wood under Marty's car last night, before the world turned wet, and I pull some sticks out and start a fire for coffee. Marty's got a fancy drink, Café Vienna — "Café Enviro-Fuck" Peacock would call it, in desperation, if he were trying to re-member the name. (Once, when he was giving me directions to his house in Tucson, he tried to remember the name of one of the side streets — it was Broomtail — and he kept groping for the name: "Two words, something to do with horses, with a horse and then part of a horse. Oh, shit, I don't know, Bridlefuck, something like that.") The Café Vienna gives small comfort, like tea, as we stand in the cold rain, and Marty looks over at his car, the albatross of his life.

Grunting and cursing, sullen, Doug arises looking as if the world's tricked him. Maybe he's wondering why he's brought a writer and an artist with a busted car and a wild-scented fumarole dog into the woods with him when he could have been alone, crawling on all fours through heavy jungle. We stand around the fire like a defeated army, with one empty bottle of Wild Turkey for each of us. To have felt so alive the night before, it feels like rainy death this morning.

We eat a breakfast of cheese, donuts, and Café Vienna. Peacock rouses himself from his morning trance to bring out his tsamba. "Eat this stuff naked and you'll shit nine to fifteen times a day," he says. Then, because it cannot be avoided, and because we're not going anywhere — those red blood cells of ours — we turn and study the Rabbit, deciding we may as well evaluate the damage. I say "we," meaning Peacock. It occurs to me that already he might feel as though he's in the Twilight Zone — like he'll never get into the woods.

We jack up the car and pull the wheel off. We stack flat rocks under the suspension to support the car in case it slides off the jack. The interior mysteries of the wheel are soon revealed to us as Peacock digs with his blunt fingers deep into the black bearing grease, pulling out twisted shards of bright metal — bits and pieces showing through the slick grease — as if digging shrapnel from a wound. Soon his arms are black up to his elbows. The newspapers we've spread out on the ground for him to sit on have become a tattered mess in the rain. There's a wheel-bearing cap that we've got to pull off to get to the root of the problem. If we can get that cap off we've got it licked, as I understand it. We have only to pop the old part out and put new bearings in, once we get them, but the cap won't budge.

We put a pair of locking pliers on it, but that only threatens to strip the threads. And then somehow we *do* strip the threads. The pliers slip, Peacock's hand slams against the black-smeared jagged axle end, and there's suddenly a stream of bright red mixing with the grease. Peacock says nothing, only grimaces, ducks his head, and looks away. I feel sure he is thinking of the Boy Scout camp-outs of his youth in Michigan, and how now, as a grown man, he should be able to get in and out of the woods without the gauntlet of responsibility for others.

We stick a screwdriver in there against that nut, or cap, or whatever it is, and tap away at it for half an hour. We need something with more leverage, more force. A rock is found, and for an hour or so we pound with that. The woods echo with the ringing sound of man's static progress, but the nut will not budge. A car stops out on the road and parks by the creek. We see through the bushes two hikers, a gray-haired man and a young woman, get out wearing bright blue rain gear, dry and happy, determined to go for a hike. They're coming our way.

"Shit," Peacock says. His shirt is open at the chest. His arms are black. He's got grease or a metal splinter in his eye, and he's blinking repeatedly. He turns his back on all of us. "No way am I going to deal with those fuckers," he says, "no way. Tell them to go on," he says in my general direction. He walks over to the smoldering fire and pretends to be entranced with it.

The hiking couple stop when they emerge from the bushes. They flinch with horror when they see our cars parked in the woods and Marty bent over the axle of his Rabbit, pounding it with a rock. The man takes the woman — his daughter? — by the arm. They glance in Peacock's direction and take in, in an instant, his motionless hunched back, and skirt our camp by a good hundred feet, not even bothering to exchange a "Hello, fine morning." When they pass Doug relaxes, glancing around with a large-eyed "Are they gone?" look. After a while we get used once again to the steady rain, which is beginning to soothe us. We seem to melt into it and give ourselves up to it, greasy and muddy, with nowhere to be.

The hardest thing about going into the woods is the transition from town to wilderness. I've seen Peacock happy as a monk in downtown Tucson, in downtown Kalispell, in downtown Salt Lake City, and I've seen him happy as a monk high in the mountains, in the clouds. It's just in between that he fidgets.

Peacock theorizes that any grizzlies that remain in the San Juans have had to alter their behavior radically. They're not yet genetically different from other grizzlies, but have taken on some of the behavior patterns of black bears, and in other ways have invented entirely new behaviors. Whereas once it was almost a given that upon emerging from hibernation grizzlies would go straight down to the river bottoms for the early spring vegetation (where they inevitably crossed paths with humans and their livestock), the remaining grizzlies now may look longingly down on all that green, spring-tasty vegetation, but have learned, over the generations, to stay out of that country.

That part of the general public, scientists included, who believe that bears are like any other animal, that bears don't think, don't have intelligence, are missing the boat. Any creature that can ride a motorcycle in the circus and learn to dress itself is capable of learning other things, too.

Peacock believes that the San Juan bears, besides having become almost strictly nocturnal, have adapted to a drastically reduced home range. He believes there are old grizzlies in the San Juans, born to those last few shot-up and stabbed-up mothers in the fifties, sixties, and seventies, who have lived nearly all of their lives in a small core area. The females, Peacock proposes, find a good hiding spot — rough country and jungle — and a good feeding spot, and they stay there, and no one ever sees them. Perhaps they adopt a higher-than-usual vegetarian diet. Perhaps, out of loneliness, they attempt to mate occasionally with a black bear, although grizzlies and black bears can't interbreed. Who knows?

We need a metric wrench or some damn thing. Peacock has various tools scattered throughout the back of his car, beneath backpacks and bags of chili peppers, but they're all the American variety, not metric. Many are left over from his job thirty years ago as a lookout in a Forest Service fire tower.

"A river to the people," Peacock says, describing how he thought of himself in those days, "a river to the people." Back then he felt it was his mission to return the government's possessions to the people whose taxes had paid for them. Whenever people wandered up to his fire tower, lost far back in the woods, he would press everything he could on them — *returning* the stuff to them was how he thought of it. If someone had a broken-down truck twenty miles away, Peacock would hand over a whole toolbox, perhaps, and a sleeping bag, and a compass and canteen, and send the bewildered pilgrim back into the woods. It is a generosity that he retains today.

I take my turn and pound on the wheel for a while. It's a Sunday.

"We're going to have to go back to town," Peacock says after a couple of hours. "Well, let's do it."

"Jeez, I'm sorry, guys," Marty says.

Peacock, with gruff tact, absolves Marty, lays all the blame on machines, on tinny cars, and Marty feels better. We find space for ourselves in the Subaru to make the journey into town. Marty manages to tuck himself into the back with the panting Keetina. Apparently she's been eating that dead deer in addition to rolling in it, and she pants blasts of gaseous carrion into the front seat. It seems that surely the fumes will mix with the chili-pepper dust and cause a fiery explosion. We drive the hour and a half down, then up, then across the Divide and into the town of Del Norte, a high plains sort of abandoned-looking place. It's so out of the way and high up in the mountains that it doesn't really feel like a town at all, but we manage to find a hardware store that's open until noon. We have ten minutes to shop, but all they've got is American tools.

We drift, searching. There's an IGA grocery store, and out of desperation we go inside, hoping against hope that in some small corner against the wall where they sell motor oil and STP there'll be one of those $5.99 cast-aluminum sprocket sets, a metric one. No luck. There aren't any wrenches, but the steaks look good. We buy a few rib-eyes and some potatoes to go with them.

I carry the steaks like books as we wander. Off in the windy distance we spy a Chevron station, the maw of its double garage doors indicating that it's open. As we draw near, walking into the dusty cold wind like gunslingers, the wind tasting delicious, tasting of snow, Peacock crouches slightly. He begins walking slower and then takes his knit cap off and runs his hand through the thin long hair he's got left. He stops in the middle of the empty road and begins swatting himself in the head. We're still a couple hundred yards away from the station.

"You guys go on," he says. "I don't feel like talking to anyone. You can do it." He makes a motion with his hands, nudging us along. "Just get a cup of packing grease. *Packing grease,*" he repeats, slowly. "Tell him it's for packing wheel bearings. Go on."

It's like some kind of test, a grail we're after. All sorts of things can go wrong. What if the mechanic asks us what kind of packing grease? Is another kind of packing grease required for the back wheels than for the front? What if we have to make a decision about things mechanical and Peacock's not there?

He stands in the street watching us leave, but when we get to the station and look back, he's gone. There's a bar at the other end of town, and a junkyard; he could be in either place. Or he could be hiding, though I've no idea where, or why.

The man who runs the gas station is acting spooked, the way an animal does when you want it to do something that you know it doesn't want to do, and you are trying to keep a secret from the animal, which is already suspicious. It's definitely like a case of the difficulty of communicating between species, and it might be more than that — mystical stuff, stuff we humans don't know about. Good karma/bad karma, charged negative ions rubbing around our heads, glimmering wild ions brushing our shoulders, dancing like glitter, invisible but perhaps somehow *scented* by this mechanic.

Marty and I are a little spooked ourselves, but with luck we're able to communicate a vague, unspoken promise of reconciliation — *we only want to give you money.* As the mechanic slowly grows accustomed to our presence, he even makes a few Sunday-afternoon phone calls, trying to find the man who runs the NAPA parts store in town.

All my life, when I'm out of the woods, seems like this narrow hit-or-miss — as if I'm on a kind of divide, scrabbling to stay atop it. If I slide off on one side, people will all be incredible buttholes, but if I can hang on long enough and slide down the other side, they'll be incredibly nice. We've definitely avoided the rocky cirque, the avalanche side of the equation. We've tumbled into the green pasture below, slid down into mercy. The NAPA man is going to wait for the end of the first

quarter — the Broncos are losing to somebody — and then he's going to come down and open his shop. He'll sell us the wrench we need
to get those damn nuts off the wheel and, if he's got them, new
bearings for Marty's Wunderwagen. It's a small store, and the chances
are not good that he will have the bearings we need, but he's going
to try.

It's been too long a wait for Peacock. Marty and I can somehow
sense his mounting restlessness. And here he comes, storming across the
parking lot. I want to tell him that it's not necessary, when he blows in
through the station's front door, all blustery with the cold air, red-eyed,
his hair wild. The mechanic spooks again, and Peacock, never one to
beat around the bush, glares at us. We didn't get the grease quickly
enough, I think.

Peacock rousts the mechanic from the phone and tells him what's
needed — packing grease — and how much he'll need. It's like a bank
holdup. The mechanic scurries into the garage. Peacock finds an empty
Big Gulp cup and tells the mechanic to fill it.

"How much will that be?" Peacock wants to know.

"Five dollars," the mechanic says. Peacock stares at him for a second.
I think he's going to give the grease back to the frightened man, but
Marty reaches for his billfold, pays the mechanic, and we're off, walking
back down the empty road with our grease. I can hear the ions tinkling
like frost crystals, dancing all around us.

Mist has formed over the mountaintops. More snow is falling, perhaps, up at the very top, but it's raining down low, scrubbing the aspen
leaves and cleaning the air we're breathing. Each step, each second
we're breathing this new air, is getting us ready for our trip, and it's
pleasant to think that our bodies are changing rapidly, that we're becoming new men.

There's time for a beer at that bar at the other end of town. We can
watch for the NAPA man through the window. We drink one round
and watch the football game on the overhead TV. The quarter is ending and we're having our second beer when we see a car pull into
the NAPA parking lot. It's raining outside, it's nice and warm in the

bar, and the waitress is friendly. Marty and I offer to go make the necessary purchases, but Peacock looks at us like we're crazy. We all go.

The man has the bearings we need, and the wrench, too. Now all we need is a cheat bar or pry bar with which to try and break that weld of the old spun bearings. There's that junkyard on the way out of town that looks promising, though it's closed.

We don't want to use up all our jangling luck at once. You've definitely got to be careful with it. You've got to hold on, and hold on, and hold on, and *then* slide a little. So, it's time to go back across the street and drink more beer and watch more football. The NAPA man won't take a tip. He absolutely refuses to take any extra money and instead begins telling us about the Broncos, about how they're so bad this year that he puts a paper bag over his head sometimes when he watches them.

We're being taken care of. It's not a bad feeling. Carpenters are playing pool at the table behind us. The light's dim and blue in the bar, not from cigarette smoke but from the rain on the windows and the blue mountains all around and the fog. The mist over the mountaintops has descended on the place. There are a lot of old people in the bar. Some of them are asleep, as if preparing for winter, while others are holding conversations so earnest that it could be the last day of their lives. Doug is growing calmer, gentling like a horse. Or like Doug.

Several rounds later — the Broncos have begun to self-destruct in a manner that makes it impossible even for their diehard fans to watch them — one of the oldest of the old men comes up to Peacock and claps a hand on his shoulder and starts talking about how he was in three wars. The guy's seventy-five if he's a day. He says his three wars were World War II and Korea.

"That's two," Peacock says. "What's the third one?"

The old guy's a little looped. "Right," he says, trying to clear his head. "Two. It was my son who was in the third one. Vietnam. That's right. I got that last one mixed up. It was him, not me." The old guy's wearing a Vietnam cap. He tells us that his son didn't make it back.

When we tell our new acquaintances that we're going into the mountains to look for bears, an odd energy fills the bar for a moment.

The air's gotten different, as if it's stopped raining, and there seems to be a feeling of pride all around, because we've come so far to see their bears. They tell us that sure, there are probably bears still in those mountains, but then the good light air leaves the bar, and they tell us that we won't find any grizzlies. No one sees them anymore.

There's no rum left behind the bar; Doug and I drank it all. Marty, who's had only a couple of beers so he can be the driver, takes the keys. We say our goodbyes and go out into the rain.

On the way out of town we stop off at the junkyard. It doesn't look as if anyone's home. We knock on the tarpaper shack in the yard — a light is on inside — but there's no answer. It's gotten colder and windier. A piss-ant black and white dog such as you might thread onto a fishhook squirms out from beneath the shack and rushes at our ankles, yapping and nipping. I can feel those ions sparkling and jangling again. It's raining pretty steadily.

We stroll through the junkyard with our heads down, hands in our pockets, water dripping off the brims of our baseball caps, and we prod various rusted assemblages with the tips of our boots. I know we're being watched, and Peacock and Marty know it, too.

Finally Peacock finds a length of pipe of the right size. He bends down to tie his bootlace, and when he stands up, the pipe is inside his pant leg. Now he has to walk with a slight limp.

We don't break for the car yet. Instead, we stroll around some more, as if still looking for the mystery part. The junkman is nowhere in sight.

"All right," Doug says, and we start back to the car. I feel as if we're stealing the junkman's most prized possession. Why won't he come out?

As we get into the car and drive away, the door of the shack opens and a man sticks his head out and yells something. He's got a pistol in his hand, and he waves it at us as he shouts.

"Chickenshit," Doug says. "He couldn't just come out and talk to us while we were there." Peacock's eyes grow small and his neck seems to be swelling, his shoulders hunching up around him. Marty's driving, looking in the rear-view mirror. Peacock tosses the pipe in the back seat. He's blinking violently, and I know that he's puzzling over the junkman, trying to imagine the motive for such bizarre behavior. The ways of humans seem so random and complex that it feels good to be getting out of town.

*

That night, we settle ourselves around another small campfire, with steaks cooking on the grill and a drizzle striking our shoulders. Keetina is nearby, her lingering fumes a reminder of where she's been. Doug looks up at the mountains that lie across the valley — the mountains so close that even the fog and the rain clouds can't obscure them — and says, "They saw the mountains but could not reach them."

Marty's car lies disassembled and broken, muddy and oily. There are locknuts in the mud, parts of the wheel wrapped in newspaper in the front seat, parts on logs by the campfire, and parts still fused to the wheel.

Tonight's story, told to us by Doug, is about Mexico, and about the writer Edward Abbey's burial place, and about Mexican grizzlies and lobos. The drug lords down in Mexico may actually be helping the lobos, Doug says, because the drug operations are taking over ranchers' turf and running the ranchers off, so the ranchers aren't poisoning the wolves anymore.

Peacock thinks there may be one grizzly left in the Sierra Madre. He found a big cleared area and tracks where a big bear had rolled in the dust. The area was much too large and violently disturbed for it to have been a black bear's. He says he talked to some of the villagers down there and they used their words for "big bear," *oso grande.*

When you get down to only one remaining bear — or two or five or ten — over several thousand square miles of country, and a good bit of that country, as in the San Juans, is up above ten thousand feet — and when the animal's that smart and that persecuted — you won't catch one of these last grizzlies in a trap. You probably won't see one even if you're out looking. The best hope is for finding a den or tracks. It's not quite like trying to prove God's existence — if there's mud or snow, a grizzly at least will leave tracks — but beyond that, it's almost the same.

There have been a few rumors, people who've been way out in the wilderness, sitting on a log eating lunch or meditating, but the odds are against you if you go out hunting the bear. Still, that's how it happened to Marty and others. The bear simply revealed itself. Marty was camped below a glacier, he says, just looking out at a cirque and the meadow below it, when he saw a dark shape come over a far treeless ridge right before dark. The next morning the steep meadow had some big torn-up spots in it that Marty said he had not noticed the day before. He didn't have time to hike across the drainage to look at them — he had to hike out that day and was already running late — but when he got back to Boulder, he wrote Peacock and told him about what he saw. And Doug began to scheme and muse.

At dawn we're back at work on the Rabbit, pulling, twisting, and prying. Threads are stripped and knuckles scraped, but finally we get the rest of the pieces off. I'm not sure of the nomenclature, but something didn't happen right. Where two pieces were supposed to come out separately, only one came out. It might have been the old bearings. Maybe the newest trouble was that we couldn't pop them out, once we got past those damn locknuts, in order to put the new Sunday-bought ones in.

It's a cool fall morning, damp and milk-foggy, and beyond the luminous fog there is an intimation of blue. We might have a decent day ahead, maybe even a lovely day, with sun and yellow leaves. Keetina, made frisky by the weather change, is racing through camp, barking and farting great wheezing deer-carrion farts. With those wheel bearings locked in there, which no amount of sledgehammering can pop out, our only choices are to make another trip back into town or to head into the woods today, a day early.

*

Peacock's fairly running up the trail. We're climbing out of the valley, packs creaking as we switchback through giant ponderosa pines, moving from patches of warm sun to cool shade, brightness to shadow, hiking quickly up the mountain as if everything that lies below, the road and the guest ranch at the trailhead where we parked, is the greatest enemy.

I should feel like we are starting out on a hunt. And for about three minutes I was able to sustain that false notion, but as soon as sweat began to dampen my forehead, the truth broke like a fever, and by the first switchback I knew that we had ourselves fooled.

The second the woods closed in behind us, our alliances shifted. Now we are going into the woods so that we can be *with* the bears, not after them.

One of the reasons Peacock is walking so fast is that he hates trails. It's as if every step on a trail is a humiliation. He longs to bolt from it and disappear into the brush, but we're back down to seven or eight thousand feet, and the trail is the quickest, shortest way up.

We won't find grizzlies at this elevation, either on or off the trail, so the object is to get out of this low country as soon as possible. But Peacock is not immune to the beauty of the lowlands. As his shirt dampens from the fast pace, he begins to cheer up, and points out, like some hurry-up nature guide, the various mushrooms we're passing.

"*Russula,*" he tells us, pointing to a big orange one. He watches the woods more eagerly as we move on through the shade. "Coral fungus."

We hear a fluttering, rustling, clucking noise in the brush above us. "Blue grouse," he says, hearing the sound first, and he raises his hand for us to stop. The large slate-colored bird appears, strutting through the pine needles ahead of us. It surveys us, then appears to murmur a worried little sound like "Oh, dear," and edges away in meandering, pigeon-toed half-circles.

The idea of meat crosses our minds, but it would mean throwing a rock at the trusting bird, and we're not feeling particularly aggressive. We want only to get deeper into the woods.

"Shaggy mane," Peacock announces, pointing to a new mushroom.

He looks down the steep slope — the valley, seen in patches through the trees, is growing smaller — finds a new mushroom, frowns (it's not edible), and says, "*Amanita*." At the next switchback he looks down at the trail, points to a cigarette butt, and says, in his same nature-guide voice, "Camel."

Our pace quickens further. I've got a six-pack of beer, as a surprise, and a couple of small bottles of Wild Turkey for campfire stories, and several gallons of water, because I do not trust the battered water filter Peacock has brought along. I stopped buying new camping equipment when I was in the Boy Scouts, more than twenty years ago. The last high-tech innovations I remember seeing were dehydrated food, which I tried once, and dome tents, which made me feel like an alien.

Cattle and sheep graze these mountains, even in the wilderness areas. We made sure not to begin our trip until after they'd been taken off their summer range, the high lush meadows around ten thousand feet. But even so, I imagine that the streams will be running gold with cattle urine, and I am determined to stay healthy. It's already too much that the cows are ruining wild lands; it would be crushing to let them ruin my digestive tract in the bargain. I'll drink my own water, thanks. Peacock's annoyed at my paranoia. He's had giardiasis before ("There are worse things," he says), and living in desert country as he does, he's become comfortable with his little blue water filter.

How can we think that grizzlies and other wild animals have changed when there are members of our own species who cannot budge? But if there are grizzlies left in the San Juans, that is what they have done — changed.

John Spinks is an assistant regional director of the U.S. Fish and Wildlife Service, the agency whose job it is to recover endangered species. "We have more grizzly work now than we can afford, so this group would have to come up with some doggone substantial information," Spinks has said — speaking of Peacock, Bruce and Lucy Bailey, and Betty Feazel — "before we'd get involved. At present, the San Juans are on our list of potential evaluation areas for grizzlies, but it's really a back-back-burner project."

We're hauling ourselves up the trail. "I hate these no-good trails," Peacock mutters, glancing up at the cliffs we're ascending. He seems liable to break from the trail at any moment and begin, literally, climbing the walls.

He frowns whenever I express reservations about his water filter. Surely there's no screen fine enough to weed out every individual bacterium.

"You could lower this thing into a shallow hoofprint of elephant piss," he says, "and it would filter out pure water." He blinks at me, annoyed at all the water I'm carrying, but I can't help it. I would rather exhaust myself lugging water than stay crouched over the crapper, gutted by dysentery.

We stop to cool ourselves in the shade. A ground squirrel races across the path, and Keetina's genes propel her to give chase. A hawk soars over the valley's terminus, and we take turns watching it through Peacock's huge field glasses. They date from the First World War and weigh about eight pounds. He's been wearing them around his neck the whole way, like a radio collar, but I have to admit they bring that hawk in close. They're battered and dented — Peacock says that when he gets in trouble he can step back and swing them in a circle and keep people away. I reexamine some of the dents after he tells me this, then look back at the hawk.

The trees are getting smaller the higher we go. We shift our packs and begin walking quickly. Once we step out of the shade on the south slope of the mountain, it is hot and bright. Even after three days of rain, it's dusty already. "Trails," Doug laments again, looking straight up into the sun. "Fucking lack of imagination."

The huge binoculars swing from Doug's neck like the bell on a cow as he leads us into one of the state's "official" wildernesses. "I feel bad that he's come all this way and doesn't like it," Marty says to me. "It makes me feel bad for Colorado."

"He likes it," I say. "He likes it or he wouldn't be here. He just wants it to be better."

Striking across a green meadow, we wade through thin air. Instead of going to the spot where Ed Wiseman, the guide, killed the grizzly in 1979, we will prowl some other cliffs and cirques — the north-facing slopes.

Peacock takes big strides; he's almost jogging. He keeps several lengths ahead of us, and I feel as if he's running from us. He's got two maps out and is glancing at them both as he hikes. The maps were published twenty years apart and drawn to different scales, and we're in the seam, the area of overlap between the maps where six or seven trails intersect.

"Fuck it!" Peacock says, thrashing the maps around and then shoving them into his pack. We sit down to rest on a fallen log. The place looks like a Fourth of July picnic site. There's the feeling that at any second other backpackers are going to burst into view.

We guzzle water. Peacock watches me drain my first quart canteen. Clearly, he's wondering how many I've got left. Sweat pools on his face, but he's beginning to relax, I can tell. There's a small cool breeze we can detect if we sit very still. If I squint at Keetina and shut one eye to block out her blue saddle pack, she looks like a wolf. Marty has her carrying her own food in her pack.

We sit on the log for fifteen minutes and let the valley below separate itself from the meadow and country above. The difference in densities is shocking. Down low, where we camped, everything was frantic and thick, asphaltic. My heart and my head felt muddy there. That little Volkswagen with the seized-up bearings, for instance — it dominated me. And that pistol-waving honyock at the junkyard occupied far too much storage space in my brain. Up top, though, things seem lighter, more vaporous. I can feel my red blood cells clicking along, tumbling into place, inflating, holding air, doing their job.

Doug finishes the last of his canteen, grunts, and wipes water from his face with his arm. "No fucking trails," he says with relief. We rise, shoulder our packs, leave the radial spur of trail intersections, and move straight out across the bumpy meadow, headed for the far tree line of wind-stunted firs. A creek runs through the middle of the meadow, quick and blue and shallow, and when we come to it we stop again, and I get to watch the great water filter in action.

Marty and Doug sit on rocks by the fast-running water. I see hoof marks on either side of the creek where cows have crossed. Like a man drawing water from a well, Doug pushes and pulls the filter's plunger, sucking water out of Cow Creek and into his fist-size filter. The water trickles out of the filter, through a second hose, and into his canteen. To me the water doesn't *sound* any different, any cleaner. My apparent ignorance piques Doug. I will hold out as long as I can with the water I have left. Plus the beer and whiskey, as well as a few cans of fruit juice.

Peacock, who was a medic in Vietnam, now gives us a technical explanation of how to beat altitude sickness, a new theory that is too complex for me to understand. The only thing I grasp about it is that it involves drinking lots of fluids. Water is best, but there's nothing wrong with a little alcohol. He has a medical theory, too, regarding alcohol consumption: one should drink five to ten glasses of water for every glass of hard liquor. Antidote chasers.

It's possible that all the water goes to his calves, bloating them to their present Herculean dimensions. His thighs are merely large, strong looking, but it's the calves that astound you. They seem massive enough to clank together when he walks, which helps explain his slightly bow-legged gait.

*

About that Wiseman bear: Wiseman is reported to have said that he stabbed the grizzly in the neck, you'll remember, with an arrow he drew from his quiver as the grizzly held him in a "bear hug."

The autopsy performed by the Colorado Division of Wildlife showed that the bear died from a wound between the second and third ribs, from an arrow passing through the arch of the aorta. Also, that wound — in the heart-lung area that archers shoot for — was circular rather than jagged. This indicates that the rib shot — the heart-lung shot — could have been caused by an arrow fired from a bow.

There was, of course, a stab wound in the bear's neck. But how did the arrow get fired broadside into the grizzly's ribs? Which wound came first? Did she have cubs with her?

The Division of Wildlife wrecked two helicopters trying to get the carcass out.

In a macabre way, there was some good news in the killing. It was the only kind of proof Peacock feels will be accepted by unbelievers: a dead bear. The U.S. Fish and Wildlife Service's Grizzly Bear Recovery Plan calls for possible restoration sites anywhere there have been confirmed sightings since 1976. Without Wiseman's "confirmation," the South San Juans would have been ruled out. Now they're in. Possibly.

Bruce Bailey, Betty Feazel's son-in-law, is loaning his services to the National Wildlife Federation. Just one bear is all that's needed for the Endangered Species Act to kick in (the act should have been enforced a long time ago), and commercial development and use of the area would have to be managed more prudently. Ski resorts, for instance, would have to be reconsidered, as would clear-cutting and overgrazing and road building.

It strikes me that these days, the best way for the grizzlies to help themselves would be to reveal themselves — not to surrender, but merely to step out of the woods and into the meadow, to come out of hiding and allow themselves to be recorded, even briefly, and then step back into the woods. In other words, a way to help them continue being grizzlies would be for them to stop being grizzlies for a moment. But nature doesn't traffic in that kind of compromise.

*

We're planning to camp at a place called Big Fish Lake, up around eleven thousand feet. The maps show that the lake is perched above a north-facing cliff overlooking the Zulu River. From Big Fish Lake, the next day, we'll follow the rim of that cliff around to one of the tributaries of the Zulu. We'll go up past a glacier and into the woods where Marty says he saw the stand of dead trees that looked as if a bear had been ripping at them, and where he saw the digs.

We come to a high meadow at the top of the world, but it is composed of low green grass, cropped by the nibbling of lambs and sheep. We would eat a lamb for sure if we found one, and how we expect the grizzlies to exercise more discipline than we, I don't know.

There are a number of trails radiating across the meadow, pounded down and rutted, and we stride over them into a thick dark forest filled with dead and dying pine trees. We climb over windfallen trees and make all kinds of noise, breaking limbs and branches. It's hot, hard work clambering over the downed timber, straddling log after log and crawling under some of the higher windfalls. We're bearing toward Big Fish Lake, which is not where the map shows it to be, but that's all right, it feels good not to be on the map.

We find Big Fish Lake before dark. It *is* large, and the lake's dark waters are boiling with trout. Peacock cackles and drops his pack. He walks the perimeter of the lake over and over, wanting to dip his hands in, I think, and splash out one of the fat trout. We haven't brought fishing rods or tackle with us. Forty yards out, the trout roll savagely, as if feeding on an insect hatch. Keetina whines and paces the shore. We split up and meander around the bank, searching for litter along the marshy shoreline and among the trees, looking for where fishermen have left refuse, tangled lines, maybe a bent and rusting hook. We check the limbs of trees to see where there may be rotting, dangling monofilament. We'll take anything.

We find nothing. The fish leap and showboat, growing larger, it seems, as we watch — inflating, perhaps, with cockiness.

I've brought a little can of Vienna sausages. We'll mix them with the tsamba for dinner, share the last bottle of Wild Turkey, and go to bed early. I try desperately, each time I go into the mountains, to carry as heavy a pack as I can, because I know I will be old any year now, and unable to carry such loads. I want to do everything, see everything, taste and smell and imagine everything, and I know that Marty and Doug feel the same way. There'll be plenty of time for a light pack when I get older.

Marty and I pitch our tents in the dimming light. Doug has moved about a quarter mile north of us and sets up his tent among the rocks at the edge of the woods, where the cliff face plunges straight into the river canyon. If you were in a big hurry to get out, you could try to work your way down through the cracks, to the river far below, and follow it out.

Marty and I gather wood. I can see Peacock silhouetted against the last red light of the mountains, battling a huge wind, trying to assemble his tent right near the cliff. The mountain winds, the end-of-day raging convective currents, set his tent flapping, clacking the aluminum poles together, and the gusts threaten to turn the tent into a hang glider. I picture him holding on fiercely as he floats out over the Zulu River, crying out with a thrilled joy like Dr. Strangelove. More likely, though, he would remain silent and squinting, shading his eyes with one hand and peering into the dimming gloom at the next hanging valley, trying to make out if those two dark shapes on the hillside are moving, squinting hard as the light fades and the cold winds carry him out and down.

We make a small fire in an old fire ring we find in a clearing in the woods. We're behind a rock ridge that separates us from the howling canyon winds. Through the trees, we can see the stars and the moon reflecting on what we're now calling No Fish Lake. The trout have stilled themselves, evidently having no interest in taunting us beyond the daylight. There never was a hatch going on; they were rolling as if scornful at our presence. We begin to pool our resources for supper, for stone soup. Marty's got some rice, and Peacock stares at me with a look of betrayal when I proudly produce my Vienna sausages, the only meat in the woods save for the grouse and the trout.

"It's been twenty years since I've eaten a fucking Vienna sausage," Doug says, shaking his head and staring at the open can of little pink pig parts. "I swore — *swore* — I'd never eat them again. We were holed up for six weeks in an isolated camp during the monsoons, just laying low, waiting on some damn thing, and that was all we had to eat. We'd catch rats and kill them and put them in a stew with the Vienna sausages, and by the fourth week all anyone wanted to eat was rat." Doug stares at the can as if it were responsible for the Vietnam War.

Sometimes I have incredible good luck. While hunting, I seem to come across the animal I'm after at just the right time — the last day of the season, with the freezer empty. I hope I have the same luck in the sprawl of the San Juans. I hope that luck — the shimmer of ions peculiar to each of us — will join our paths with those of the grizzlies.

In the meantime, I aim to learn things, mostly about bears. And to hide. What I like to do in these mountains, in this cold clear high country, sandwiched in that thin wedge of fall between the end-of-summer backpackers and the coming hunters, is hike up into the rocks, into the forests and meadows, into the past, and hide from the increasing chaos of civilization. Of which, when I return, I am a part.

But I do not feel part of civilization up here, walking lightly in the cruel harbor of these rocks, walking along the edge of the coming winter. If I am careful and move with respect through these mountains, I can separate myself from my civilized life, or at least the odor of belonging. In the depths of the San Juans, and in other mountains, are places where this ability to go in and shed one's coat still exists, even if only tenuously.

Native people believe that bears can communicate with the spirit world. Bears move in and out of the earth, disappearing underground during winter's hibernation, and therefore belong to two worlds.

Doug's staring at those Viennas like a wrestler, as if he's preparing to charge in and defeat them. "Okay," he says, "I guess twenty years is long enough. I guess I can eat them."

We cut the sausages into buttons and drop them into the simmering tsamba soup. We let Keetina lick the gelatinous sauce in which they're pickled.

"Rat-head stew," Peacock says, horrified, shaking his head. It is our species' eerie fascination with the past, our magnetism for it, that causes us to repeat it: the way it's always springing up in our faces, just enough to make us believe that there is a pattern, controlling it from above and below as we skate across the surface. The past, where we've come from, to which we will always return.

We make peace with the warm beer I've hauled up, and chase it with hot Wild Turkey shooters. The world below us is unlit. There's not a road or cabin in sight.

We're enjoying each other's company. But at night, alone, looking up at the stars, there's that hunger, that empty spot.

I roll over on my side as if to listen to the interior of the mountain, as if trying to hear some clue, one way or the other, yes or no. But I can't hear a thing, just my own pulse.

We wake to a hard frost. Stirring the night's ashes, about to build a new fire for breakfast, we find among the coals small metal ferrules from a fishing rod, and nearby a few snapped and frayed fiberglass-rod fragments — signs of some fisherman gone berserk.

"The lake of broken dreams!" Doug howls. "Lookit," he says, picking up the evidence. "It wasn't enough for the guy to break the rod over his knee. He had to try to burn it."

We find more pieces of rod in the bushes, similarly mangled.

"Ghost fish," Peacock says. He's jubilant, picturing the men without fish breaking and burning their equipment, while beyond them trout leap and splash, never touching lures, flies, or hooks. Wild trout. "Poor fuckers," Peacock says, piecing the carnage together like an archaeologist.

There's tsamba for breakfast, and coffee. Peering through the trees, we see steam rising from the lake and a band of butter-colored elk, eight cows and calves, stepping cautiously through the meadow hoarfrost like a stealthy war party. At the lake they take water, then move on into the trees. Later we hear a bull's flute-like squeal.

The fish that will not allow themselves to be caught in the lake of broken dreams are a good omen. We're cold, we're moving slowly, but things feel good. The heat of the lower world is being shed bit by bit. It's been more than twenty years since Peacock was last in these mountains, mapping outcrops for the University of Michigan. There were grizzlies up here then, though no one knew it.

If bears are in the San Juans now, they must be more secretive than any society of bears ever was, more secretive than any society of people. Perhaps the grizzlies cover their mouths with their paws on cold mornings, to muffle the mist clouds rising from their nostrils, their warm lungs, to avoid giving themselves away.

Almost one hundred years of hiding, like some great and plagued biblical tribe. According to several accounts, Theodore Roosevelt's included, even early in the century the grizzly was uncommon in Colorado. John Murray, in *Wildlife in Peril: The Endangered Mammals of Colorado,* writes of "the last grizzly bear killed on Lone Cone Mountain in San Miguel County on May 26, 1907; a big bear, with four and one-half inch foreclaws. Sheep and cows were being funnelled into the high meadows, and the ranchers were pouring hot lead into the big bears, eliminating the boldest ones first; eliminating the bold ones completely."

*

To our way of thinking, solitude means altitude and thin air. We'll keep climbing and look for digs and other activity right at tree line and in the meadows above tree line, expecting bears to venture into the appealing meadows and grasses under cover of night. It's the way we prefer to distance ourselves from human activity — to climb to the top of the mountain — as do other wild creatures, such as elk and bighorn sheep and mule deer. So many things with a wild heart operate in this way.

Peacock's blunt, crooked finger traces the topo map, showing the verdant side canyon a few miles up, where Marty saw the claw marks and the fresh-dug earth. Our day lies ahead of us, and maybe nothing more. What if the bears aren't here? What if they've vanished from the earth, from history?

Have you seen the snow leopard? No! Isn't it wonderful?

Keetina catches a mole, trots back with it limp in her jaws, and Marty's proud enough to burst.

How many layers must we take off, how much of the other part of our lives must we shed, before the bears will show themselves?

*

After breakfast, we walk along the rim of a gorge, look across, and wonder: Are there any bears over there? If not, then maybe a little farther west, a little deeper in? We weave back and forth through the woods at the edge of the gorge, heading down elk trails, searching for a bear's summer fur that's rubbed off on the rough, sappy bark of fir and spruce trees.

We search for scat and tracks. We move slowly with our packs, getting adjusted not so much to the great altitude anymore, but to the feel of these woods, this country. Elk are bugling along the river far below. The aspens blaze on the rocky cliffsides across the gorge. The river rushes straight through the valley floor, white spilling rapids. There's no meandering, no lazy oxbows.

We're headed toward the one bend the water does make, where it strikes the fault of a cliff face. This is our goal for today, where Miasma Canyon empties into Perdido Canyon. Who knows what magic, what unknown geomagnetic spirit lingers along that fault scarp? Perhaps the spirit still glides up and out as the fault, with geology's creep, settles and shifts, gives off spirit gas and shapes the land and our place on it.

In these mountains there was a grizzly known as Old Mose, who was seen and tracked for thirty-five years. Old Mose was reputed to have killed two or three men in addition to the rich diet of sheep and cattle that ballooned his weight up to around twelve hundred pounds. He reportedly measured over ten feet long, with almost a ten-foot girth.

Thirty-five years would be a long life for a grizzly anywhere, and twelve hundred pounds a record for a grizzly in the lower forty-eight states. Without doubt, this ravager's genes run through any remaining San Juan grizzlies: that tenacious ability to hunt, and to defy death. Old

Mose evidently knew what it took to keep from joining up with death, from taking that underworld boat ride, but he was finally killed on April 30, 1904, John Murray writes, shortly after he left his den on a mountain. The hunter waited for a bear to come out of hibernation and shot Old Mose as he emerged into this world, stomach empty and seeking life, seeking grubs and ants, seeking hot white light.

These bears love to endure. They are so similar to and yet so different from us that perhaps it is expecting too much to believe that we'll do anything other than wage war on them.

We stop and snack often. Marty's brought some sunflower seeds. We sit on a log in the sunlight at the edge of a high-elevation bog and share a green apple. Keetina sits below us in the marsh grass, in the shade of the log, and pants, looking precisely like a wolf. A hundred years is not a very long time. That's all the environmental movement is playing for these days — sanctuaries for a few remnant wild places. You can't go back in time a hundred years, or even twenty, with man's place upon the land. You can read about it and dream it, but the sad and simple truth is that often we're not trying to change the world, nor even save it, but instead merely to endure.

These days, we're trying to protect the last unaltered places and keep them connected — or reconnect them in some fashion, so that wildness can still travel through them, like electricity.

In 1691 Henry Kelsey, an explorer in America for the Hudson's Bay Company, was the first white man to record an encounter with a grizzly bear. "To day we pitcht to ye outtermost edge of ye woods. . . . Nothing but short round sticky grass and buffalo and a great sort of a bear which is bigger than my white bear . . . and silver hair'd."

John Murray writes, "Kelsey relates that, after killing a grizzly bear, he found the meat good but was discouraged from keeping the hide of the bear by the Indians because they said it was God."

*

Through the centuries, the riparian meadows below us lured the San Juan grizzlies out of the mountains each spring, down to the season's first green-up and to the sheep and cattle and homesteads. And it was the same for wolves. Those green meadows in the spring, so tempting, so necessary.

Only those who learned to turn their back survived. Those who did not seek to dominate, but only to hole up, carve out a place, and endure.

Doug and Marty and I finish our lunch, sit in the sun a bit longer. Are we making this search for the bear or for ourselves? There is a huge difference, I know that much. A man is not a bear, and a bear is not a man, but there is definitely a mixing going on, a merging. Some days it feels like this search is for the bear, and other days it feels like it is for people, for the heart of our own species. We could just as well be searching for a cure to one small kind of cancer, or for a lost hymn. We could be up on the Canadian line, going over into the Canadian Rockies. We could be down in Mexico, in the Sierra del Nido, the Sierra Madre. Why here? Why in the middle of the country, the square-ass middle of the Rockies, dead center between Houston and Kalispell? It has something to do with a centering, with the concept of heart.

*

It's a good year for blue grouse. Their populations fluctuate in cycles, and surely this year is a peak. They're huge and strutting like turkeys in the shadows ahead of us, moving nervously and making small *putt-putt* noises. Marty often has to hold Keetina back on her leash.

Sometimes we flush the dusky blue birds when we slip between the trees and onto a grassy hill where they are sunning themselves. In those moments I find myself wishing for a shotgun. A classic pass shot as they set their big wings and glide down the mountain, and I'd have dinner for three. Doug stares after them wistfully and says the same thing, how a shotgun would bring dinner, though he admits he hasn't hunted since Vietnam. "I can't be around people with guns," he says firmly, this being one of the war's smaller prices.

The end of bird hunting is not the end of the world, of course. You adapt — you adapt or fracture. You hole up and hold on to that which has not been lost. You reduce your home range from five hundred square miles and a cocky I-own-all-I-survey attitude of strength to a range of two hundred and fifty square miles and a don't-bother-me philosophy. Then to a range of one hundred square miles and an I'd-rather-be-alone philosophy. Then to twenty miles, and then to four or five. Instead of chasing off others of your kind, as you might once have done, you gather in and band together, you hold on.

The shrinking wild habitat of the soul.

It's hot, passing through patches of sun, and we stop again to drink water. I now trust Peacock's filter and am given small comfort by the knowledge that if I catch giardiasis or some coliform bacteria, the symptoms won't manifest themselves this week.

We move across the high plateau, the gentle top of the mountain whose meadows turn the sheep so quickly into bundles of fat summer cash. We try to avoid the rutted crisscrossing livestock trails, but it's hard. Anytime we come to a sizable meadow, they're there: cow pies and short grass and deep ruts with the black earth showing, as if motorcycles have been racing down the trails.

Peacock grunts through gritted teeth. "It's a shame," he says. "They have done a fucking top-notch job in screwing up this meadow. It took a lot of fucking *energy* to hurt it this bad." The world is turning against him, the friction of it throwing sparks in its counterspin as it tries to halt his costly, inefficient philosophies of loyalty, land stewardship, and acknowledgment of a spiritual presence in the world.

Where else is there to go after the land? To the blue sky? To memory?

*

When a grizzly cannot control its five hundred square miles, it reduces its range to an area that it can control, in order to survive. Instinct, once so important, atrophies; forethought and individual characteristics — eccentricities, individual traits and talents — become magnified and significant. The individual becomes more important than ever. Old game laws tumble out of the science books.

They're here somewhere. I'm starting to feel it as I look across to the farther, rockier mountains. There are places where our pale pipestem legs, our ribby backs, cannot take us without days and days of journey, and there are places where cattle and sheep surely cannot go. These bears have taken Old Mose's legacy of endurance and power and, in dwindling to a remnant population, have combined his raw strength and fury with a developing individualism. It is like the birth of a new species: the thinking bears of the San Juans. Murray writes of Old Mose in *Wildlife in Peril:* "The brain weighed fifteen ounces, just a tiny fraction of the gross body weight. This is about the weight of the brain of a normal newborn human baby. The centers of smelling and hearing were . . . enormously developed. The optic nerve was small, and the optic regions at the rear of the brain were poorly developed. . . . The brain was wide across the areas controlling motor activity."

*

We share another snack deep in the forest, at the edge of a tiny mud hole where there should be bear tracks. We sit on fallen trees and pass around a bag of raw tsamba. My legs are weak from hauling my pack on this spartan diet. Already, my inflatable stomach, normally stretched tight and full with whatever I desire that day, lies empty, shrunken somewhere inside my rib cage. There's only sun-dappled forest, clean air, and the tiny shining meadow to give nourishment. I realize I haven't been truly hungry in a dozen years.

The river lies two thousand feet below. We're high on a forested island of the mountain. We rest our legs for a while, staring trancelike as if our eyes could will a bear to come out of the shadows. I feel strongly that sometime in the past — twenty, fifty, seventy years ago — a bear must have passed through the meadow for this spot to mesmerize us so. The meadow would have been a small lake then, the bear walking casually along its shores.

Eventually we shake the spell, the earth telling our bodies what our minds cannot. We move on, heading for the west-facing cliff that overlooks Miasma Canyon, the spot where Marty thought he saw fresh digs on his previous trip — places where a grizzly had been ripping the earth, looking for food — and then, just before dark, two large shapes coming out of the forest, across the canyon.

Our canteens are empty as we hike toward our next camp, down game trails and over the lovely-smelling sun-dried pine needles of the dry forest. Our canteens are still empty hours later, when we arrive at the dolomite outcrops of the sheer cliff and look to the left, upriver, at Marty's hanging green valley, which already in midafternoon is half shuttered in shadow and gauze-like light, an impressionist painting.

We're all three slick with sweat, even in the hard, cold wind that's coming off the hanging valley. The thin river shines blue far below us. One of us could make a hero's journey to fill our empty canteens, half a mile down the cliff and out to the river and then back, but that would alert any bears to our presence. We're able instead to find enough water to moisten our lips by scrabbling out onto the dolomite outcrops and sucking small puddles of rainwater from the occasional depression. We keep our bodies low to avoid broadcasting our whereabouts — re-membering, after a couple of days in the woods, that the existence of one thing is tied to everything else, and that to stay hidden from the bears involves staying hidden from deer, hawks, voles . . . all of it linked, all of it shying, understandably, from human contact.

"Why don't you and Marty bushwhack back to Clearwater Lake and get us some water," Peacock suggests, studying the map. Clearwater is a high, narrow lake lying to the east, perhaps half a mile through the woods. Marty and I understand that Doug wants to be alone. Although his demeanor suggests nothing, we can feel the desperate ions clanging around inside him, can almost hear their echo, so we leave him to scan the meadow below by himself. Marty and I take a compass bearing and head straight for where the map tells us the lake lies. Without the map, you'd never believe there was a lake on top of this flat mountain.

There's a wind white-capping waves on Clearwater Lake when we come to its shore, feeling like Lewis and Clark. The waves splash across the rocks at the water's edge, as Marty and I find comfortable seats before beginning the awkward trickle-squirt pumping of the filter. He holds the canteens in place while I pull and plunge the piston, and

Keetina gulps her water greedily, directly from the lake. Wind ruffles her fur. She stands in the tall grass beside us, then turns and trots along the shore.

"I wish they'd bring it all back," Marty says. "Wolves, buffalo, grizzlies."

We sit at the edge of Clearwater Lake, filtering out pure water a few drops at a time.

C oming back through the woods in an increasing dullness of light, I feel like a hunter, and it's a good feeling. Am I cutting blood from a turnip, flaying the pig's ear, in stating that as the human world revs up to its most furious, frenetic complexity there is a great downshifting, calming distillate to be found in reducing one's being to the pursuit of the simplest tasks?

Finding water. Drinking it. Walking through the woods at dusk with a friend and his wolf-dog.

When we return to where we started from, we find Peacock gone. He's moved camp. Nothing remains of a human presence but a small conical pile of rocks.

We ease down the trail, searching for his earth-colored tent and his startling fluorescent pack. As might be imagined, Peacock will never be called a clothes horse, although his friends at Patagonia generously outfit him with top-of-the-line equipment for his wanderings. He's talismanic, treasuring small magical objects he's found, such as arrowheads, bear's teeth, gemstones, and feathers, but his joy in touching or "owning" these objects is spiritual, not material. The only piece of clothing he gives a damn about is Ed Abbey's old suede sportcoat, which Peacock wears only on special occasions. He likes the fancy new equipment, but in milder months one is likely to find him wearing whatever clothes he happened to pull out of his closet: side-split, fire-welded hiking boots, a mechanic's oil-stained and torn khaki pants, and a brilliant mango-colored sweater to top it off. He has various hats that protect his head from the sun.

To some extent, the fancy camping equipment embarrasses him, makes him feel like a fantailed guppy, I suspect, sent into the woods for breeding or ornamental purposes. Although he's proud of his lurid new backpack, with all its secret compartments and functions, he doesn't quite know what to do with all that color. Sporting bold colors makes climbers and their equipment highly visible, even in fog or blizzards, which is a boon to rescuers and lost or injured mountaineers. But Peacock's need for deep hiding and deep peace is so prevailing, so without compromise, that thoughts of rescue fail to clutter his mind. He even carries with him a dropcloth with which to camouflage his pack if a plane should pass overhead: a white sheet for snow country, an olive-and-drab camo cloth for the other seasons. This is not to say that he wouldn't know what to do in an emergency. As a former medic and leader of desert search-and-rescue teams, he knows how to keep alive not just himself, but others.

*

We find him farther upvalley, toward the mouth of the cirque — a head-high rock spike shielding him from the valley's view. He seems happy to see us and waves with great animation. We were gone a long time, and it occurs to me that he might have been worried for us. Something familiar jars my mind from long ago, and as I look down the cliff at Peacock, who is still fifty or sixty yards away, I remember what it is. In Edward Abbey's novel *The Monkey Wrench Gang,* the character of George Washington Hayduke, which is based on the real-life Peacock, is standing on a cliff in the desert when two members of the gang first encounter him.

> Smith fiddled with his field glasses, looking for something he thought he had seen moving on a distant promontory above the gorge. He found his target. Adjusting the focus, he made out, a mile away through the haze of twilight, the shape of a blue jeep half concealed beneath a

pedestal rock. He saw the flicker of a small campfire. A thing moved at the edge of the field. He turned the glasses slightly and saw the figure of a man, short and hairy and broad and naked. The naked man held a can of beer in one hand; with the other hand he held field glasses to his eyes, just like Smith. He was looking directly at Smith. The two men studied each other for a while through 7x35 binocular lenses, which do not blink. Smith raised his hand in a cautious wave. The other man raised his can of beer as an answering salute.

"What are you looking at?" the girl asked.

"Some kind of skinned tourist."

*

We spread out behind boulders at the edge of the cliff and look down at the green meadow far below, willing the great bear to come sauntering out as night's protection approaches. It's windy, and a raven glides across the valley, floats down the cliff, passing by not ten yards above our heads, fixing us with a cock of his head and an eager, conspiratorial look. The look he gives us, passing so close, has something to do with that notion of our hearts being locked. The raven is like a master jeweler or a locksmith, one who is also searching, if not for a key, then perhaps for a freshly killed deer. He glides by, never once moving his wings, like a spirit, drifting.

"Mister Raven," Doug says quietly, and salutes him. A minute later, after the raven's gone, we hear its caw, like laughter.

The sky was filled with ravens on the windy day we met in the red-rock Utah desert in the spring of 1989. We were at Edward Abbey's memorial service, north of Moab. Doug had been Abbey's closest friend, and had helped tend to Abbey when he was dying. Doug was now tending to him for the journey to the hereafter. When I first heard Doug speak, I sat on the slickrock with thousands of other people and listened to the tributes. A blue-juniper haze extended over the red rock of the plateau for as far as we could see. Ravens everywhere, and loss.

*

A little later, just as the sun is sliding behind the crags, behind the lunar scree, I hear a ripping sound like the hissing fuse of a twenty-dollar fireworks rocket, and a small raptor shoots straight up the cliff face right in front of me, spinning twice in tight quick circles before flying away fast — a peregrine falcon, a creature only slightly less endangered than the grizzly. It is the best of omens, yet darkness falls without bears in the meadow. The wind is furious, as if the rest of the world has somehow tilted during Doug's absence from it and all the wind in the world is now pouring along the ridge. We make some tsamba stew over a sputtering fire in the rocks, trying to block the wind, and divide a can of Vienna sausages three ways. Hard stars glitter by the millions, each possessing the brightness of soul of a person who lived before us. We are tired and retreat to our sleeping bags early. I can feel the mountain, huge and coiled and dense beneath me, as I sleep lightly on top of it. The wind rips at my tent all night long.

*

I'm up at first light. The wind's rushing harder than ever, but Peacock's already out on the cliff, motionless, swathed in his sleeping bag, and I know that once again, for the moment, I am seeing the bear.

Light slowly returns to the valley. Marty crouches in the little rock shelter we've built and lights his stove after a hundred tries. We boil water for instant coffee and more damned tsamba.

"Ah," Peacock says, leaving his post and making his way toward us when he smells the coffee. "Chemicals! Chemicals!" He takes a deep breath of the cold rushing autumn air, sips the coffee that Marty's handed him, and I'm thrown into the past again, recalling *The Monkey Wrench Gang:* "He fixed and consumed his humble breakfast: tea with powdered milk. . . . Sufficient. Squatting close to the fire, he sipped his tea. Chemicals: his mind cleared."

It will be important to me, all my life, to hear Peacock grunt "Chemicals! Chemicals!" and picture him drinking his morning coffee. A sense of tradition is one of the grooves in the lock along which the key must slide.

The bears in the San Juans have tradition, too. We can feel it — the fact that whatever shred or tatter of their species is left up here is a traditional one, steeped in the culture of their hard-surviving ancestors. It's a matriarchal society, and the mother's havens and trails and habits have for the young bears a lucid significance. *This is where you run when you hide. This is where you do not go. It is all right to eat berries from this bush. This stream, but only this stream, is all right to drink from. Beneath this tree is where we will nap, our family, for as long as we are on the earth. Only this tree, in this drainage.*

Our persecution has made these bears incredibly smart and strong, but we have discouraged their spontaneity, their freedom. If only we could loosen the constricting bands around them, perhaps our own hidden wounds, our own limits to spontaneity, would begin to heal. We have lost these grizzlies and lost our relationship to them. We have lost a part of ourselves, of who we were and who we will be.

Do we really need another ski resort, yet another road? Wouldn't it be nice instead to have something new in our lives, something our forebears had but then lost?

Run when you scent humans, or their food, or their livestock. Run away.

I am not arguing for the bears, because that cannot really be done. They seem beyond argument, like whales or clouds. What I am arguing for is a little space for the bears. Again, it is like a lock, a system of intricate grooves into which a key will fit. Grizzlies in Colorado? The lock is aching to be opened.

We pick our way along the cliff, heading south in the morning light, looking for a fracture, a chute down which we can slide into the valley. Everything's vertical, and we make two false starts down cracks in the cliff (Keetina whimpering, watching from above) before we find one that might work.

It's more of a waterfall, a notch out of which, midway down, a small spring is seeping. Ferns and red-leafed currant bushes grow thick on either side of the chimney. Doug goes first.

The cliff is composed of knotty conglomerate rock, petrified mud flows from half a million years ago, and so there are round knobs that our hands can clutch and tiny niches where we can wedge our heels and toes. We make our way down the wet, shadowy, cool chimney like spiders. All I can see below me is Peacock's bald head descending into the tangle of currants: he's half sliding and half climbing. To let go of your hold in the chimney would be to plunge three hundred feet to the rocks below, at the first break in gradient, and after that what was left of you would bounce and roll the rest of the way. I dislodge a small rock and it clatters down too close to Peacock's head before sailing away.

"No more of that," he says without looking up. "It's slippery."

Above me, Marty is having trouble with Keetina, who refuses to enter the chimney. He tries to coax her into following him on a leash, but she only barks and then howls, mournful that the earth is swallowing us up. Finally Marty climbs back to the top and gathers Keetina under one arm. She struggles and scrabbles, and he resumes his descent.

About midway down we all begin to absorb the sickly, treacherous knowledge that we can't go any farther down but can't go back up. It's really slick — greasy-green-algae slick, with a quarter inch of water trickling over everything.

The chimney flanges to a three-foot ledge — the head of a new waterfall — and then disappears below. There's no way to tell what lies under that lip, over which the water is running. We gather at the top of the waterfall, clutching our faint toeholds and handholds like bats, breathing hard. Keetina, still struggling in Marty's arms, is breathing hardest.

We can see the valley, but we can't see what the next drop in the chimney is like. Another small safety ledge might be only ten or twelve feet down, or then again, there might be nothing.

I find myself considering the curious fact that misanthropes like Peacock always seem to place themselves in situations that define their existence in modern society — dead-end routes with no outlet of escape, no resolution but death — but the fierce pull of gravity, preternaturally stronger by the second, chases these thoughts away, and I dig deeper into the clay with my fingernails.

There seems to be an anvil writhing in my backpack. Then I realize it's my heart pounding, telling me to go down, to go over, but every other part of me is trying to stay stuck to the wet slippery cliff above the falls. With my back to the cliff, I imagine that even my buttocks are trying to squeeze tightly some small irregularity in the rock; that everything and yet nothing, only my will, is holding me above rocky death. Worse yet, it feels as if the slippery algae I'm pressed against are beginning to release me, as if the algae are secreting a chemical that makes the surface even slicker, which will cause me to slide loose and be swept down over the falls.

Peacock's slipping off his pack, looking angry at being trapped, and he hands it to me, then slides butt-first over the edge. We hear a thud and mild cursing, rocks scattering and rolling and bouncing to the bottom. Doug grunts and curses again and we know he is all right, that at the very worst he has broken his ankle.

I edge along the moss wall, closer to the edge.

"It's not too far," Doug shouts up to us. He's impatient, anxious to hit the green meadow, the valley below, now that the excitement is over. "You can ease that pack down to me." He's standing on another small ledge about twenty feet below. I crouch and hang his mud-smeared pack over the edge. We don't have a single rope between us.

"Ah shit, just drop it," he says, and I do. He slips when it lands, then scrambles to cover the pack with his body, like after a fumble. "Now your packs," he says.

Ever the sissy, I say, "We could toss our sleeping bags down first and use them for cushions to land on if we fall."

Peacock looks aside, furious, then growls, "All right," and examines his left forearm. There's a long crescent-shaped gash in it, his blood mixing with the mud and algae. I unstrap the sleeping bags, toss them down to him. I make a chain of the bungee cords that held our sleeping bags to our pack frames and lower our packs to him with the cords.

The packs still won't reach him, but now it's less of a drop, and Peacock uses his body to break their final fall. The empty Wild Turkey bottles clink when he catches my pack, and for a moment his eyes light up with the hope that I'm saving a surprise, but I shake my head no, and he laughs.

Now it is Keetina's turn. She's still wearing her blue saddle pack full of dog food. Marty and I knot the bungee cords through and around her pack, with her still in it, to form a harness. Her yips and howls echo off the cliff as we lower her scrabbling over the edge, and then she's treading sky, swimming in midair. Marty and I dig in and hold on as Keetina yo-yos up and down. It's as if we've caught the biggest fish in the world. She struggles and spins, dangling just above Peacock's outstretched arms.

"Drop her," Peacock says. "I'll catch her."

Marty's face is ashen, and he looks physically incapable of releasing the cords that hold his sweet dog. But she's too heavy and writhing too much to pull back up. Same old story: when there is no choice, the only choice is down.

"He'll catch her," I tell Marty. "It'll be okay." Marty's face grows paler. "On the count of three," I say. My shoulders and hands are tired, and I'm afraid Keetina is going to pull me over the cliff with her. She's air-paddling harder, as if she knows what's coming.

"One . . . two . . . three," I say, and we let go of everything. The bungee cords whip away from us like slingshots. Too late we realize they are going to snap Keetina, and snap her hard. She yelps and there is a clatter, and when Marty and I peer over the ferns, Peacock is lying on his back, inches from the cliff, and he has Keetina wrapped up in his arms. She's quivering, whining, lunging to break free in her panic, but he's not letting her go anywhere. Like a paratrooper, Marty vaults over the edge of the falls, lands on the rocks below, and stumbles over to comfort his dog.

It's crowded on the small ledge. With packs and people strewn everywhere, it looks as if a light plane has crashed. I half slide, half scrape my way over the edge, gripping with my fingernails for as long as I can until I, too, am falling.

From here on, everything will be easy. It's still a vertical chute — "A first descent!" Peacock crows — but there are handholds once more, and there's the good knowledge that each step we take means we have that much less distance to fall. The chute begins to develop a slope near the bottom, and there's a new jungle of vegetation. We have descended into a hidden valley that could not be more perfect for grizzlies.

"Cow parsnip," Peacock points out, waving his hands over a lush, broad-leafed, pale green, juicy plant. "*Heracleum*. They don't have to dig for it. It's right here."

We're almost to the bottom. Peacock points back up to the top of the cliff, the reef over which we spilled and trickled, like blood. "I found a couple of digs along the cliff last night," he says, "but they sure as hell weren't fresh."

Then we are into the valley. Like Marty, it's my last day, too. I must get back home, and I don't think Doug's going to let me head out alone. His sense of loyalty is a thing he went into the army with (the kind they wish they could teach) and a thing he will leave life with. At least we made it to the place we wanted to reach.

We cross the river and hike toward the cirque. To our left timber spills over a saddle that leads in and out of the valley, and that's where the bears may be resting in the daytime. It's also where Marty found the tree that he believes was marked by a grizzly's claws. The tree wasn't on a trail, but he thinks he can find it again.

It may be very simple. If we can find some silver-tipped fur where the bear rubbed and clawed the tree, we'll have our proof. Our search will be over shortly after it started. We'll have found, in less than a week, what the state of Colorado has not been able to find in the eleven years since the Wiseman grizzly was killed.

It's a little scary, that responsibility. Peacock is adamant about the future — that there be no live-trapping of these bears, no radio-collaring, no helicopter-buzzing, no pestering. We're all hoping that if we can go in and prove the bear's existence where others have failed, then our recommendations will carry not just increased validity, but a kind of jurisdiction. That is, if we can show we're correct in our belief that there is a grizzly population in the San Juans, then our recommendations for the management of that population will also be correct: a hands-off policy will be more important for preserving grizzlies than any other policy. If we find the bears — *when* we find them — we must turn our backs and walk away. We must give them more space and quiet around the edges of their territory, more of a buffer, and then hold our breath and hope they make it, hope they can expand and prosper in the same way we put tiny twigs on coals and try to coax flames from them.

We know this will be the hardest thing in the world for a professionally trained biologist to do, with all those computers and all that electronic equipment at his or her disposal. Poison voodoo darts and stuff like that. We've got to make them see that what's left, this remnant, is not a few lost straggling bewildered animals whose lives are in need of saving by any agency, but instead a tight society of bears that have learned to leave man alone. All the grizzlies need is wilderness, the last breath of it.

We're climbing to the headwaters of the river. The meadow's lovely, with grasses and cow parsnip waist-high in places. The green cirque is small and perfectly shaped, reminding me of a scoop of lime sherbet. Above the cirque lies talus, huge tumbledown quartzite and dolomite boulders. Above that, the shining bare cliffs, capped at the rim with ovals of snow, and more alpine meadow. Higher, at the top of the cliff, there is tundra. It looks like the country that grizzlies in Alaska roam fearlessly, digging roots, chasing squirrels.

We sit on a huge boulder, a glacial erratic, that holds the three of us, the still quivering Keetina, and our packs, with room to spare. We spread out tsamba for lunch. It's a cold, bright day. Elk paths meander through the rich high meadow, the grasses bent where the elk passed this morning. The wind makes a soughing sound. We are a long way from roads, from a cabin, from anything of human design. We are about as far back in the present as we can get here in southern Colorado.

We sit in the cold sun and glance back at the chimney we descended and at the trickling blue stream, the tilted meadow, and the sheer dark wall towering above us. The only possible thought one can have under these conditions is *it's good to be alive.* Perhaps thinking this is what reminds Peacock of when he worked with a search-and-rescue team in southern Utah, for he begins telling stories of death, of fatal mistakes, as we sit in the sun and savor life. I find an orange at the bottom of my pack, and we split it three ways.

We go off in different directions, wandering the meadow like children, looking for scat. It's a long shot. Scats with grass in them form the best samples, while scats of other composition — berries, meat — tend to decompose faster. But we move through the meadow anyway, with our heads down. Black bears grow large in these mountains, so we won't be able to lean as heavily on the biologists' rule of thumb (not that Peacock ever leans on rules of thumb of any kind) that grizzly scat comprises a quart or more of waste matter. What *will* provide us with a near lock on a grizzly is the presence of silver-tipped hairs. There's a pathologist at a crime lab who examines DNA, and if we can get a hair sample to him, he can find out what we've got, black bear or grizzly. He's examined dozens of samples from all over the West, and he's about got a fix on it. Bear scat might contain the hair we need. Grizzlies groom themselves often, and in so doing lick loose hair from their body and sometimes swallow it. The hair passes through them undigested and unharmed.

We move through the blowing grass like detectives. This is bear heaven, at high altitude. But we find only elk droppings, and we're disheartened. We tell ourselves, "Well, if there are only two or three or five of them left, the odds are next to impossible of finding one of their old shits, especially in this harsh, quick-changing country, with storms every afternoon in the fall." We tell ourselves, "Well, hell, everybody knows bears shit in the woods. They're not like cattle." And we even tell ourselves, "It is now the nature of these intelligent creatures to hide evidence of their presence."

Peacock says that grizzlies are aware that they leave tracks, that hunted bears will avoid stepping in mud and will instead move from rock to rock. The bears' (and humans') cousin the raccoon will even cover its eyes with its paws when hunters approach in the night with searchlights.

The enormity of our task rises before us like a wave, and our optimism waxes and wanes. One or two bears in all of the San Juans? We're disappointed. This high meadow is perfect. We can only guess the trouble is that it's too close to sheep and too close — three days of hard hiking — to the sign of people. We look across the mountains and tell ourselves — knowing we're right, and sad about it — that the bears, if they're out there, lie farther, deeper in.

As we leave the meadow, angling toward the woods, I look down and spot a neoprene camouflage glove resting on the tall grass like a severed hand. It is a bow hunter's glove. The tight-fitting neoprene allows the hunter to pull back the bowstring while keeping his fingers warm on a cold morning. I know instantly that this hideous outstretched hand, balanced on the grass as if placed there this morning, will spook Doug no end.

I'm spooked myself. The grass is unbent around the glove; no trails radiate from it. I can't help but stare. Peacock's coming up behind me, looking at the blue sky, basking in it, and I know I need to turn away or snatch the glove up and hide it. Suddenly I realize how it got there. A hunter must have been circling this dead-end ski bowl of a meadow in a plane or a helicopter, and when he leaned out to get a better look at the elk, his glove fell and floated to earth.

I turn away just as Peacock approaches. Marty is already across the river and up at the woods line, anxious to go find the bear tree. Peacock *smells* the glove, I think, before he sees it. He wrinkles his face and stares at it.

"You dropped a glove," he says, pointing.

And I have to tell him "No, I didn't." I look at him as if to say, We did not get far enough in.

There needs to be places where any of us can get far enough in, even if only in our minds. We should not have to travel to the Himalayas or to the Congo to find such a place. We need to have interiors in this country, at least a few for each state, places against which to measure your spirit, as Wallace Stegner has said. And in Colorado, the San Juans are the best candidate to become such a place.

"Cock motherfucking sucker," Peacock says slowly. "Greasy anus-wipe dicksuck fuckwads." He is looking my way, and I'm afraid that *I'm* going to have to answer for this glove. I know he doesn't see me. All he sees is a mask of red webbing, a mist of blood, but it's a terrible look, and I wish to hell he hadn't come across the glove.

It's part of an old argument: what constitutes wilderness legally, and what constitutes it morally? I say the definition is simple: wilderness is a place where wild things live. "No motorized vehicles" is one of the congressional parameters for wilderness, but isn't a helicopter a motorized vehicle? Is it truly wilderness if hunting guides come in a week before the season to count and delineate — and while they're at it, do a little hazing, a little herding of the elk? Doug calls this a two-dimensional wilderness, a place where machines can still hover and flit and roar only a few feet off the ground.

Peacock and Marty and I are not against elk hunting or flying airplanes. We do both. All we want here is one place — *one place* — where the grizzlies can live. One wilderness. There wouldn't be a need to stop elk hunting, just those damn overflights.

We stare at the bow hunter's glove. Peacock has run out of curses.

Marty's wondering what we're staring at. He knows we've found something significant, but knows better than to call out.

It doesn't take a mystic or a prophet to figure out that this trip has entered a downhill slide. Bad karma has definitely set in. We have pressed too hard against the thin membrane separating the burden of history from the fragility of the present. We can feel the momentum shift as hugely as if the trip were mounted on rusty hinges, as if an iron door were creaking shut.

I can't leave the glove lying there, littering the San Juans, but I'm hesitant to touch the thing, to put the heat of its symbolism in my pack, close to my body. But finally, because the feeling of the wild is gone, I pick up the glove and stuff it into my pack.

We hurry across the creek and up into the woods. It's a fine old forest, big Douglas firs and ponderosa pines, with cedars along the feeder creek that trickles off one of the cliff's glaciers. There is hundreds of years' worth of soft duff underfoot, and a game trail, the kind bears like to pad up and down. The slant of light is pleasing, as is the sound of the creek, but Peacock is as spooked as a bull elk, a stag with the hounds closing in. He's charging up the hill and off the trail at an alarming pace, his bare shins scraping against branches, but he says nothing. I can hear the small branches snapping as they poke his shins, can see the blood from his legs trickling down. I like to think these wounds allow some of his rage and feelings of betrayal to bleed off, the rage like a gas in his blood which, if left unvented, would expand until it blew him to smithereens in front of our eyes.

"I'm sure it was somewhere around here," Marty says, wandering from tree to tree, looking for the claw marks. Peacock is tight-lipped, silent, still bashing his legs against things, tripping and stumbling, and I can tell that he wants to get the hell out, but this is what we came for, to investigate this one tree in the forest, if we can find it. As soon as we find that tree, I know what the game plan will be.

We will hike until we are out of the mountains. We won't stop. The air has turned bad. We'll not be a part of the valley where the glove was dropped, not today. It is going to take a lot of winter snow to cleanse that image from our minds.

We follow a game trail up toward the saddle, and then Marty decides the tree can't be this far up. He descends again, and Peacock's smashing over branches again, completely flustered, trapped by fate and by this century. He is on the verge of proclaiming all of this a grand fuck-up of a place and bolting for home.

As we weave our way down toward the creek, we begin to notice small blazes on some of the larger trees. Then we see some honyock's name carved in the scarred-over bark of a tree — *Joe* — and realize we are following an old sheepherder's trail: maybe back in the twenties or thirties, he'd come down from Wyoming, ride through here on his way between valleys, checking on his sheep.

While I think these pastoral thoughts, Peacock says, "Grizzly-killing son of a bitch," cursing the dead man's soul. It is not a bright day for the hopes of reclaiming the grizzlies' center, their *querencia* in southwest Colorado.

Beside the creek — Marty has now decided we've come too far down, that the bear tree is higher up in the woods after all — we come upon a frightful scene, as if we've descended into the underworld. Skeletons of martens and fishers dangle from rusted traps that are nailed to the trunks of trees — an abandoned trapline from so long ago that the trees have sealed over lengths of the chains with their outer bark. The bleached marten skeletons are intact, protected in these heavy woods from storms. They hang there with their snapped legs and silent, screaming mouths and empty eye sockets.

"This is bullshit," Peacock yells. "We've got to get out of here." He's quivering with the wrongness of the world and the way the day has changed, but just then, on the hill above us, Marty finds the tree.

Before we take a look, Peacock and I go from trap to trap, trying to pull the traps down, but they're too welded to the trees. We grunt and tug at them with both hands, even climb halfway up the tree and pull with our hands while pushing against the trunk with our feet. I even swing from one of the traps, hang my dead weight from it and jerk my body downward, but neither of us can undo what's taken a tree decades to do. We have to leave the traps where they are, though we remove the skeletons from each trap and return them at last to the earth.

We flee the swampy creek bottom, striking back up the hill to where Marty and Keetina are standing. It's a relief to have a different sort of focus. We circle the tree slowly, like boxers. It is a big tree, half fallen, leaning against another, and it occurs to me that Marty must be about half wild himself to be able to find this one tree again.

There are four gouges in it that could have been made by bears, but the tree's in such rough shape — it's beginning to rot — that the long marks could have been caused by the rotting process itself. The gouges are down low, equally spaced, and each is the thickness of a large claw, but up higher the bark's all feathered and rotten, and it is hard to read what's going on. The only thing for certain is that it's a very old tree.

"We've got to go," Peacock says.

We exit the woods, skirt the meadow on the side of the creek opposite where the archer's glove was found, and strike down the valley almost at a jog, headed for the south fork of the river, which we will follow for ten miles before it crosses under a dirt road. From there, one of us will travel the dirt road up to where we left our car at the guest ranch. It'll be about twenty miles in all, and it's already early afternoon. (Peacock has retained the soldier's knack of always knowing what time it is, despite not owning a watch; "Twelve-thirty, twelve-forty?" he guesses, and he's right.)

We tramp down the canyon through marsh and glen, still looking for tracks, but with something inside us injured. We stop where the creek enters a heavy swamp, a jungled knoll that slows our descent, and because the weather's turned windier and colder, we take a quick tsamba break, making a soup of it on Marty's stove.

Rather, we attempt to make a soup of tsamba. Just as the water is about to boil — we're seated on driftwood around the hissing blue flame; Keetina's down at the river, drinking — the stove explodes, sending skyward the pot and stove shrapnel and a shower of sticky gruel.

We're all three drenched with hot tsamba. Peacock stands up — there are pieces of the muck in his beard — and without saying a word he shoulders his pack and strikes off into the woods, angling for a shortcut that will get us to the south fork faster.

Marty and I gather up the stove pieces and the cookware, load our packs, and hurry after him. He's already out of sight, but we can see bushes and saplings shaking where he has entered the jungle. We can hear his grunts and curses. Keetina pauses at the forest's edge and growls, raising her hackles, and we enter the thick woods following the blood sign.

An hour later we find a talus gorge to clamber down. We don't even have to carry Keetina this time, and we are back in the lower meadows. By the middle of the afternoon we have once again reached the network of livestock trails with their ankle-deep hard-packed ruts that cut down through the green grass and into the black earth in a spidery, aimless fashion that reminds me of the most tortuous of varicose veins.

Peacock is going at a good clip, and we have to trot to keep up with him. Now he stops abruptly and begins cursing like a banshee. He's stumbled upon another animal trap — this *century* is a trap for him — because to get out quickly we must take the grisly man-reeking trails.

After a while we pass a group of canvas tents, where outfitters have come in and set up their village early. If we had the time, and if our luck was running white hot instead of sour, we might investigate the liquor supply in those tents. It sounds as if I'm advocating thievery, but the tents make me feel as if *I'm* the one who's having something stolen. For an outfitter to come into a wilderness area and set up a village and then leave it vacant, like a time-share condo — it's outrageous. But we pass on without raiding their liquor.

Up and down small rises, through stands of aspen with lovely yellow and orange leaves fluttering in the breeze, bright colors against the blue sky, we move to the river's slow sound off to our right. Two mule deer pogo-bound across the trail in front of us and up an oak canyon, and again, as with the sticks jabbing and scraping Peacock's legs earlier, I can feel the violent gases in his blood seeping out, relaxing him a notch or two — as if he's giving up, a preacher might say, to the bliss of death and the opening of eternal light. Watching the deer, I can feel my own heart uncoiling, and that is what it's like.

But horseshit, fresh green horseshit, begins to pepper the trail, which becomes more deeply rutted and slippery, devastated by the passage of pack trains during and after the week's heavy rains. Up in front of us, Peacock is making small strangling noises. Our boots are muddy and shitty, and we jump back and forth across the double-rutted trail (two horses abreast, giving the trail the appearance of a road through the wilderness). We dive off the trail and bushwhack along the stony river for a few minutes, but it's too slow. We have allowed ourselves to get into the most dangerous of psychological predicaments: we're in a rush, running late.

Late to where?

We pass a campsite of hunters who are sitting in the small meadow around a roaring fire, cooking wieners on sticks and drinking in the middle of the day. Their two deer hounds, southern black-and-tan starved-looking bastards, jump up from their masters' feet and tear savagely across the meadow, barking at Peacock, who's in the lead. The hounds nip at his hamstrings.

Peacock stares straight ahead, marching furiously down the trail, unwilling to acknowledge the presence of the hunters or the dogs. The hunters are gazing drunkenly at the scene, and only when Keetina, thirty pounds lighter than the hunting dogs, rushes up and knocks one dog off the trail, leaps on the other one and grips it by the throat, about to tear its windpipe out, do the men rise from their lawn chairs. "Hey, hey," one of them protests, "cut that out!"

By the time Marty and I have the dogs untangled — one of the big black-and-tans is yelping, missing an ear, and there's blood on Keetina's muzzle — Peacock has disappeared a hundred, maybe two hundred yards up the trail. Once again we hurry after him. Marty praises his wild wolf-dog, and in the manner of life, of nature, by pressing on through the hideous, we have been rewarded with the sublime.

We find Peacock standing over a mud wallow of horse piss in the center of the rutted trail. There is all manner of hoof and hock marks here, but on top of those tracks are fresh bear tracks, lots of them. They're as fresh as tracks can get: the turned-up mud has not even oxidized.

The bears are fleeing the forest, too, as the outfitters and the black bear and elk hunters flood in. It's an exodus of bears. They must have moved out early this morning, under cover of darkness.

There are three separate sets of tracks: a big black bear, a middle-size black bear (no claw marks fore or aft), and a baby bear — perhaps two babies. Nothing to get thrilled about, yet. The black bears are all following the trail *out*.

"They know they're being hunted," Peacock says. He studies the ground fiercely. His engagement with life, his love of it, is returning. Marty and I can feel the gears meshing again, the way a transmission catches with a palpable *thunk*. Peacock is Peacock again. And that's when we find the track.

It angles across the trail, on top of all the black bears' scattered prints. It's big and shows narrow, straight-lined toe prints. Peacock freezes over this one track, which is headed in a direction nearly ninety degrees away from all the other tracks, and he crouches over it slowly.

"This could be a grizzly track," he says carefully. The size of it is what's exciting. It's nine inches long.

Peacock is staring at the track as if he did not believe he'd find a grizzly track in the San Juans. I know he believes the grizzlies are here, and if this were Yellowstone, he'd point to such a track, say "Grizzly," and pass on. But it's the context, here on the horse trail, and at a lower elevation, coupled with the strangeness of the day, that gives him pause.

"This doesn't look like a black bear track," Peacock says, walking gingerly around it. He crouches again and points out how different it is from the other tracks. The straight-lined, closer-together toes can reflect the grizzly's long claws sticking out in front of the foot (black bears' toes tend to spread apart), but we can't find any claw marks where we'd expect them because the trail has been roughed up by the black bears. Still, everything else reads like a perfect grizzly track.

We search frantically for another track, but this bear was wise and left nothing. It had come from the river bottom and angled up an oak canyon to the north, moving away from the black bears, in no way eager to cast its lot with those bears, whose tracks the hunters would find and follow, perhaps with dogs.

Peacock is frustrated that there's only the one track, but he's excited by its size and shape, and by the intelligence behind it, the ninety-degree contrariness of it — the rock-hopping awareness.

"When they know they're being hunted, they're aware of tracks," Peacock says. "Tell me that's not indicative of a higher fucking state of being." After we take a picture of the track, he smushes it out with his foot, erasing it from the hunters' view, and says, "Good luck, buddy."

We head down the trail as if rescued, each of us with a huge, bright secret. The trail is dark and shadowy, canopied in places by the cool firs and brilliant aspens. The thing that Peacock turns over and over in his mind is the surprisingly low elevation. We would have expected to find a track like this at 12,000 feet, but not here. A hundred years ago, even fifty or sixty years ago, this is where you could find grizzlies in the San Juans. But to find a present-day grizzly down this low?

Most of the nature photographs on calendars and in coffee-table books show grizzlies rollicking in meadows or in alpine tundra, but that's more a reflection of the photographer than of the bears. Thomas McNamee writes in *The Grizzly Bear* that "in Yellowstone, about eighty percent of their time during daylight hours is spent in timber, where sight lines are short."

Peacock wants badly to lean all of his scientific belief on that one track, but even in his "question science" mode, he knows he can't bet his soul on the one partial track being a grizzly's. Later, in a letter to the U.S. Fish and Wildlife Service, he will tell his heart's truth: ". . . Let me outline what we saw. Two of the three sets of tracks were black bear but the third was different: the track measured eight and three-quarters inches from heel to toe, the toe prints closer together and in a straighter line than the smaller sets. This sign, however, was a single track, and I am reluctant to pronounce judgment on a single track beyond saying I didn't think it was a black bear track."

We have made a discovery that will affect all of the rest of our time in the San Juans.

It seems, then, that the bears make their primary home range not in the highest timber but down in the second or third band of timber, among the smaller, more private side-hill parks and meadows and in the denser, wetter fir jungles. They appear to be living, when unstressed, between nine thousand and eleven thousand feet. Perhaps this strategy gives them more opportunities for escape: to hide, they can move laterally, into the jungle, or they can descend, or move higher.

After we leave that track and head down the trail, we feel altogether different, in the way that a birth or a death can make your life different. It is like holding a helium balloon and then releasing it, or like parasailing on water skis when you first begin to rise, separating from the water. You are no longer buzzing over the rough chop but are instead lifted. Even the light in the woods has changed; it seems a hundred years cleaner. Shadows at the edges of the trees seem to pulse slightly, respirating, as if a bellows in the heart of a bear is gently breathing air back into the mountains.

Marty and I hurry to keep up with Peacock. His steps are jaunty, determined, as if he's in a rush to get out and tell the world. Ten minutes ago he was wearing a festering badger's scowl, but now he's expansive, generous and free. He doesn't seem to mind so much when he turns around and catches me scribbling notes on the palm of my hand and on little scraps of paper. "Drives me fucking crazy," he says, but nothing more. Once, when I asked him why he thinks so many writers cross over into what's called nature writing, he said succinctly that nature is a way to "alleviate nightmares." Literature is about passion, he said, and it follows that writers and others are going to be passionate about objects and places of great beauty. "Nobody ever got lyrical or mournful over a busted thermostat."

We saunter down the trail, all but whistling. We can barely contain what our eyes see: the intensely blue sky and the yellow leaves of fall, the flickers of movement, birds and sometimes deer and elk back in the bushes. But Peacock's mood is acutely variable, particularly on this trip — or rather, on the part of it where he's forced to walk on a trail. More green horse turds litter our way. We're compelled to navigate around stinking puddles of foamy horse piss.

"Hunting," Peacock says. "If it's done right, with respect and ceremony, it's okay, but in the twentieth century . . ." He shakes his head like a dog with porcupine quills stuck in him, and grinds his teeth. "Basically, the greed of hunters like these brings out a whole lot of unhealthy and perverse impulses that don't belong in this ecosystem." Though Peacock has hunted before, he doesn't see why some hunters have to harm the land in the act of hunting. He turns this way and that, again catches me making notes on a little square of paper, and seems even more furious. His eyes widen to show their whites, his mouth settling into a frown — enemies! enemies everywhere! — and he accelerates his pace.

A goshawk flies through the trees and lands on a low branch, turns its head and looks back at us like an executioner. "Goshawks tend to land in the middle of trees rather than at the top," Doug says, calming once more.

The predator watches us pass, studying us, as if evaluating each of our hearts.

The afternoon stretches, and the mild sun grows cool on our backs. We're hauling ass, heads down, all sweat and muscle and one foot in front of the other, surging to make all the miles before sundown. Peacock pauses once for water (he sweats like a fountain, summer and winter, which is one reason he doesn't like cold weather; if he stops moving, the sweat freezes on him) and scowls at a trail where a pack string has taken a shortcut around a river bend, gouging a bare stripe across a meadow.

There are more empty tents set up at the far end of the meadow, flaps moving gently in the breeze, ghost tents awaiting their guests. To make matters worse, there's a Forest Service trail sign, tacked to a tree like a street sign, and once again Peacock catches me jotting down notes.

He rages at the outfitters. "They've got their own little empire," he says, "and they don't want anybody else trying to horn in on it. They want exclusive ownership of the animals in the woods so they can trap and hunt and maim."

Peacock's savage lust for the stillness of wilderness isn't solely the result of a mistake of fate, of his time spent in Vietnam. I'm always surprised at the incredible likeness of heart of all people, and I believe strongly that whatever works for Peacock in his trembling, ravaged state will work for us all. I know that places like the San Juans work for me: the strength of knowledge that no matter how chaotic things get in cities, there still exists in the wilderness a system of logic and grace. The San Juans do not exist for this reason — they were here before us and they will, of course, outlast us — but it cannot be denied or ignored that certain places can recharge and cleanse the spirit, can wash away the befouled crust of unhealthy impulses and alienations we've accumulated.

What we don't have much left of in the West is the West itself: the place to the left of the East, up and over the first wall of mountains and then down into the wild jungled basins below. I think of wild basins like that as being the healthy marrow in a bone, which produces the brave cells that keep the organism of the West alive, preventing the West from drying up, scarring over, and vanishing.

There is only one San Juan Mountains, one Yaak, one Targhee.

The southernmost known range of grizzlies in our fine, fading country is around Yellowstone. But the grizzlies here in the San Juans are closer to Lubbock than to Yellowstone, closer to Wichita and Austin and El Paso, closer to Santa Fe.

The canyon through which we're thundering is growing dark and cool with end-of-day magic. We stop downwind of a herd of elk that drifts across the trail in front of us, those lovely orange coats with the yellow butts moving slowly, proudly through the aspens. There's a small spike in with three big cows, and I call twice on the grunt tube I've brought along, trying to get an answer. The elk stop and turn toward us from the safety of those trees, staring at where they think we are — we're crouched, hidden behind a fallen ponderosa pine, its huge trunk smashed recently by lightning. When I bugle again, the elk lower their heads — the spike looking like a shaman — and trot on deeper into the woods.

As dusk settles in and the pace quickens, Peacock begins to grind his teeth and grunt and call out single words, "Hup! Whoa! Who-*ap!*"

Sometimes he twists his head violently as he barks, and at first I think he is calling out some marching cadence. I struggle to stay right behind him. Occasionally he makes a sound like a real word, and I lean forward and try to close the gap between us and ask, "What's that, Doug?" He only digs in harder, with his backpack creaking like the leather saddle on a horse. He's laboring to pull away from us but he can't quite do it: we're a step or two behind him.

Finally, at the edge of darkness, Peacock pauses for breath. He's sweating like a wild pig, as usual, though the evening is cool, almost cold. He ducks his head once as Marty and I gather around him, and he utters a single *"Beep!"* and then, *"Fuck it!"* He looks up, amused.

"I've got that Tourette's syndrome," he says, "not bad, just a little. It mostly comes out when I'm tired. I haven't been talking to you, haven't been saying real sentences. Sometimes words just come out."

The sweat's beginning to cool against his back and chest. Already, though our knees and legs are still throbbing, it's time to push on again, up a hill. We hurry into a narrow tunnel of aspen trees. The earth is black beneath our feet, rich swamp muck littered yellow with leaves. The hill slopes down toward a creek and our senses cloud: in the aspen tunnel, stepping through the kaleidoscope of black earth and yellow leaves, it's so disorienting in the gloom that we stumble and slow our march. We have to set our feet carefully, as if walking down a flight of steps in the dark. There's nothing but black and yellow flashing before our minds, registering in hot stabs against our desiccated and weary brainpans. The dimming light falls as if an accomplice of the strange meadow, the tunnel of confusion. It's as if the night is swallowing us.

In the full darkness we have to stop in this ghostly land and reorder our senses. The gold coins of leaves lie scattered beneath us, glowing dimly, and they seem to be radiating a rejuvenating power. We stand and wait for the leaves, and our minds, to cool. The blackness of night slides in over the leaves, flowing around our ankles and knees until it is safe again, and the blazing leaves have sunk into memory.

We can feel the valley only a mile or two ahead of us, can see the yawn of it, the belly of stars it's holding. We have not made it out by dark, as we hoped, but we will get out. Once we reach the gravel road that bisects the valley, there'll be another six miles to where our car is parked, but for now there is only the aim of emerging from the woods.

We cross a wooden footbridge, the moon now shining on the rolling water beneath. We pass an abandoned cabin and its fallen-in root cellar. We haven't eaten in ten hours, haven't eaten anything but tsamba for days — tsamba and a few Vienna sausages. A yearning for hot food, any food, stirs within us as we rise out of a sage plain and come upon our first barbed-wire fence. We separate the strands for one another to slide through, passing our packs over the wire like commandos reentering the world of men. Once we reach the lonely caliche road, which runs pale like a river down the center of the valley, we shed our packs and throw on warmer clothes. We're still up around nine thousand feet, and the stars flash and sparkle like hot jewels.

Because my mind's still reeling from that aspen tunnel an hour or so back, and because, although I do not feel good, neither do I feel pain, I volunteer to head down the road to pick up the car, and my offer is accepted.

It feels good to have the heavy pack off. The gravel road, illuminated by its own shining paleness, makes a good path, so I won't have to worry about stumbling. "Maybe a car or truck will come by, and you can hitch a ride," says Doug, but we all know there will be none.

I take off at a run, like a little boy. I still feel part of the bear magic, woods magic, mountain magic all around me. Perhaps when night fades I will be all man again and will have to surrender whatever it is that's in me that seems so different tonight. But morning is a long way off, and it feels good to run.

Elk stand in the middle of the road or in the sage on either side. Their dark shapes look like horses, and strangely they do not run off as I come loping down the road. They wheel aside as I approach. I can smell their warm bodies, can even smell the crushed sage on their shins, can hear the creaking of their bellies, grunts and farts, clops of hoofs. From the river, near the dark line of trees below, I can hear the bugling of one bull who dominates all others. Cows and spikes and perhaps silent bulls are trickling down out of the foothills to go find the challenger, and at the next crest in the road I stop for a moment, panting, to rest and listen.

It is pleasing to imagine that there are wolves around me, pulled in by the sound of the elk. I want to feel that the woods are involved with all parts of itself, a symphony of flesh and blood and tooth and fiber and muscle. Instead, all I hear is the one shrill note of the bull's bugle, riding straight out across the pasture as if drawn on a line, and the bull is unable to call forth any wolves, which have been gone from here for fifty years. The last wolf in Colorado was killed in the San Juans in 1943. They went fast. In 1936 there were eight known wolves in the state, but by 1938 there were only two. Then just that one, five years later. But we got it. Even in 1985, when a pet wolf escaped into the San Juans, we got that one, too.

If wolves return to Colorado, they will probably be reintroduced into Rocky Mountain National Park, where the huge elk population flirts annually with disease. John Murray, in *Wildlife in Peril,* quotes a 1976 graduate thesis by Herb Conley, of Colorado State University, who claimed that the park could sustain up to two dozen wolves — two or three full packs — without depleting the deer and elk in the park and within a twenty-five-mile radius around the park's exterior.

To make the San Juans whole, to remake and heal them, we must start at the top of the pyramid, restoring and protecting the habitat of the great predators such as the bear, wolf, lynx, and wolverine. We must also repair the base of the pyramid. The water must be kept free of cyanide, and the soil must be held in place by forests, rather than washed away by the scrubbing action of cows and sheep.

I run again, through the elk whistles, through the night. Soon the road slopes downward, increasing my speed. I feel like part of the San Juans. After thirty minutes, the dark outlines of old barns rise in the meadow below. It is the closed-up guest ranch where we left the car.

I have less than a quarter of a mile to go when headlights appear behind me, far up the road. I slow to a walk and then stop, breathing hard, enjoying the way my diaphragm and rib cage are rising and falling, drawing in cold night air, filtering it, absorbing it, then sending it back out. One thin rivulet of sweat trickles down my right temple; I can feel how chilled it gets as it goes. I know I should hide, but perhaps Marty and Doug are in the truck, and besides, it's too late: by now whoever they are have probably already seen me.

What if they're elk poachers or man killers? What if I'm the only witness to their slow-moving truck as it eases out of the mountains? My reinitiation into the world of men is not a trusting one.

An old man and an old woman ride in the battered flesh-colored truck, which is pulling a flatbed cattle trailer loaded with firewood. They stop to offer me a ride, not knowing I'm only going to the next pasture. I hop on the cattle trailer and dangle my feet over the edge, jouncing among the wood and the chain saws (there are two, and I realize that the old woman must run one of them). The night wind stirred by our passage is surprisingly cold — it's well below freezing — and after about thirty seconds we're there. I jump off the trailer, say my thank-yous, and wave goodbye. Just the right amount of human contact. Reentry by stages.

The truck rattles away, and I trot down the trail to Doug's car. When I open it, the inside reeks deliciously of green chilis. My eyes sting and water as the pepper dust swirls everywhere. I can taste it burning in my mouth, in my nose.

I turn the car around and drive back up the road with the headlights off, aiming the car by moonlight, driving with the windows rolled down. It is like swimming upstream in a cold river as the air washes over my arms and my face and fills my lungs.

*

I almost drive past Doug and Marty in my ghost car. They rise from the sage like elk and run out into the road and flag me down. I gather them up and we head for the little store in Platoro, the Skyline Lodge. The night's bitterly cold at that elevation — Platoro, an old gold-mining village turned hunting camp, is above ten thousand feet.

"You guys go in," Peacock says. We're standing under the stars, sketching our grocery list on a scrap of paper while leaning on the hood of the car. "You just go on in and get these things. Ask for a chicken and some meat, some big porterhouses, and some vodka and gin. I'm not going in with those . . . those *people*. I'm not ready to fuck with that." He shrugs as if to gather warmth.

Marty and I start up the steps. Marty says, "This is where they keep the bear that was killed in 'fifty-one — one of the last grizzlies killed in the San Juans, up until the Wiseman bear."

A sheepherder named Al Lobato killed the Platoro Lodge grizzly up near the same spot where Wiseman killed his. After Lobato killed his bear, in August 1951, the next grizzly, a female, was killed by a government trapper, in September 1952. She'd had two cubs, and both escaped. Either of those cubs could have been the Wiseman bear, twenty-seven years later, or perhaps the parents of the Wiseman bear.

Peacock is pale in the moonlight. "Shit, I forgot all about that," he says, looking up at the lodge's silhouette. "Motherfuckers," he says softly. "Go on in."

Marty and I enter the lodge as if going into enemy territory. A boy and a girl are sitting on a couch playing cards, and there are three men and a woman seated around a table next to the couch, also playing cards. They stare at us in surprise, as if they've been caught at something — they probably didn't see us drive up, since I never bothered to turn the headlights on. Marty and I must be wild-bearded and smoky-smelling and perhaps a tad pig-eyed from all the day's miles. We're still in our camouflage clothes and heavy boots.

The people in the lodge do not seem to understand that we have come into their store aiming to purchase something. Instead, they freeze, still holding their cards so that no one can see what they've got, and watch us with what seems like guilt, waiting for Marty and I to make the first move.

We break eye contact with the card players and go over to the grocery section of the store. There is a little café on the other side, but it has long since closed down for the night. Marty and I prowl the two aisles — no chicken or steaks here — and find the meaty substitutes we think will satisfy Doug: tins of sardines, saltine crackers, a block of Swiss cheese, a block of cheddar, apples, oranges, and limes. A bag of ice. Some candy bars. Pop-Tarts. We do the best we can and take our armfuls of goods up to the counter. Behind the counter is an assortment of liquor bottles.

No one comes to wait on us. The people keep playing cards, and they're not saying anything. We stand there scanning the back shelf. We clear our throats a couple of times.

The little male bear — only the head, which was sawed off right behind the ears, to hide how thin the young neck must have been — is mounted at chest height above one of the tables in the café. He's snarling, but with his ears laid back he looks more like a housecat than anything else. He was only two or three years old when he was killed, and even now, many years later, he seems to be crying out at the injustice, the utter uselessness.

We need a drink.

"Excuse me," Marty calls to the card players. "Can you sell us some vodka?"

The woman, young and pale with long black hair, looks at the two older men first, and then at the young man, who nods. She gets up and comes over to the counter. Not a word is spoken during our transaction. She pulls down the bottles we request — I throw in a bottle of cheap red wine, because the night's so cold; it makes me feel like I'm adding good rich blood to my system — and we're gathering up our things when Peacock bursts into the lodge looking so frazzled, so confrontational, that it seems certain he believes we're being held hostage.

Peacock has two empty blue plastic cups in one hand and an empty Styrofoam cup in the other. The card players stop again, and I watch with amazement as expressions of stealth and cunning come down over their faces like a sheath. *They're simply pretending we're not here,* I think, and then, *Why? Why are we frightening them?* No one speaks. Peacock, trying to assess the situation, spies the bear head on the wall and walks over to examine it.

Marty and I watch as Doug touches its teeth, its nose, and smooths back some of the fur around its eyes. From where we stand, Doug seems to be petting the creature. He might even be speaking to it. The woman leaves the counter and goes back to the card table in the middle of the store.

Marty and I drift over to the moth-eaten bear. It looks tired now, as if its spirit wants nothing more than to rest. Doug is pressing his thumbs and fingers lightly against the bear's toy-savage features, and when we draw near him he straightens up, checks to be sure we've got the groceries, and turns and hurries out.

In the empty parking lot, under the cold moon, we pour stiff drinks, adding only a faint coloring of orange juice to the vodka and ice. We sit on the hood of the car and sip the drinks, letting our fluttering hearts calm a bit. We expect to hear voices, even laughter, resume in the lodge now that we're out, but we hear nothing. Doug looks at us, blinks hard as if to reassure himself that we're on his side — *there are three of us* — and then we get back in the car and begin the trek down the mountain, through twenty miles of night and switchbacks and elk trotting back and forth across the road that will lead us to our camp.

By the time we get there, we will have covered, on foot and by car, sixty miles today. If our theories are correct, the grizzlies, on this rich, long, bright day of early fall, will have holed up and not moved very far at all — perhaps only a mile or two.

These grizzlies are not Montana or Alaska grizzlies. These grizzlies are like no others in the world, and if we lose these, in the manner that we are letting so much else rush through our fingers . . . We humans are gunning ahead too fast, leaving bits and parts of ourselves, of where we've come from, smeared all over the countryside, the wreckage of our spirit.

In the meantime, more whiskey! As the wholeness fragments, breaks down, we can take solace in individuals — our families and friends.

When we get back to our high camp, we find that Marty's car has fallen off the jack, and the injured wheel is resting in a mud puddle. We also left one of the windows down. Yellow aspen leaves have blown in, blanketing the seats and floorboard.

It's well past midnight. We build a fire, and the firelight leaps and jumps across the sunken car behind us, just at the edge of the shadows. As we haven't covered nearly as much ground as we hoped to, Doug says he is going to make a quick strike up the intriguingly named Growler Creek, near Mahatma Pass, where grizzlies have been seen in the past, and which might be one of their hole-up spots, another *querencia*.

A light rain begins to fall. As we finish off the food and the bottles grow lighter, stories get told. Marty is especially troubled by the little bear's captivity in that bad-karma lodge in Platoro. How easy it would be for someone — not us, of course — to go in and snatch the head and return it to the woods!

"I can't stand to think of that bear sitting up there forever," Marty says, throwing a log in the fire. "Their goddamn trophy."

Doug nods and looks into Marty's eyes. "I whispered some secrets in that little bear's ear," he says.

When the bottles are empty, Marty and I head off to our tents, and Doug unrolls his sleeping bag by the fire, pulls a plastic tarp over himself to hide from the rain, and yields to the day's end. The stars and the night and the rain and the mountains sweep in over us.

*

In the morning the camp is sodden and overhung with mist. A light rain is still falling, but across the valley, above the mountains, there are patches of blue. Asleep on his back with his mouth open, Doug is catching significant amounts of rain. Marty and I watch him — the water must run down his throat and into his stomach, like water in the desert flowing into a hollow rock. Finally Marty gets a long branch and, as if snapping a trap, touches Doug lightly on the shoulder with the end of it.

We're six feet away, but still we jump back when Doug sits up like a bomb going off, arms outstretched. He blinks and looks around at all the puddles of water, and at Marty, who's holding the branch like a lion tamer. Doug's old black stocking cap is wet, and he takes it off and slowly squeezes the water from it.

We get a fire going, knowing better than to try to engage in conversation, and after a quick cup of coffee and one of the powdery, awful Pop-Tarts, we break camp and start the new day, the transition day, which for Marty and me will involve broken machines, phone calls, and exchanges with clerks, cashiers, and attendants. It's hard to leave the woods and go back to a world of so many humans, but it is our world now, and has its own strange calling for us. We drop Doug off at Mahatma Pass and watch with envy as he disappears into the brush like a bloodhound. He gets the woods, we get town.

In Del Norte the town's busy, but the sky is blue. The sense of elevation is delicious. We've acclimated: our blood cells are beefed up, drunk on the wild. Right away we find a samaritan who has a vise, a ball-peen hammer, and a torch; the old fragments of Marty's wheel clatter to the garage floor after only a few blows. Five dollars.

The samaritan explains to us how we can pack and press the bearings ourselves. We leave, but soon afterward get cold feet. We go back and ask him what it would cost for him to do that job, too, and we make a wise decision and pay him. Another five dollars.

Marty and I make a few leisurely phone calls while basking in the sun, remembering how slow life is supposed to be and how hard it is to step away from the roaring pace of the world. Then we head back to our camp thinking that it wasn't so bad after all. Around two in the afternoon, I leave Marty and cruise up toward the Divide, where I am to meet Doug at two-thirty.

It's cold and windy where I park. I expect to see Doug come strolling down the ridge at any second. Tom Cartwright, the father of Clarke Abbey, Edward Abbey's wife, once told me about the time Peacock was supposed to rendezvous with the Abbeys and the Cartwrights for a Thanksgiving dinner out in southern Arizona's Cabeza Prieta badlands — a vast country of black rock, thousands of square miles of raw land that's off-limits to people because it's used as a bombing range. No bombers were flying on Thanksgiving. The Cartwrights and the Abbeys had told Peacock a week earlier that dinner would be served in the desert at three in the afternoon.

Doug was just roaming the desert, wandering from water hole to water hole. No one knew for sure if he'd be able to find the place where he was supposed to meet his friends, or if he even knew what day Thanksgiving was. Meanwhile, the turkey, wrapped in foil, was pulled from the coals, steaming and succulent. Clarke Abbey lit candles. Somebody unwrapped the foil. It was a big turkey and had been cooked slowly on mesquite. In all directions was the baked flat desert, which rolled out to the horizon and then disappeared into the haze.

Tom Cartwright thought it was time to open the wine. Someone screwed the corkscrew in and pulled slowly, carefully. There was a small *pop!*

"Look," Tom Cartwright said, pointing west.

A dark upright object, a man, came walking out of the haze a long way off — coming in from out of the shimmer, wavering, but coming in fast.

So it is with complete confidence that I am watching this sheltered ridge line. At any second, I know that Doug's going to come trotting over the top.

When a half hour passes I get out of the car. A storm is not far away. I can feel the thick, sluggish air pressure, the wind is increasing, lifting and rocking the car on its shocks, and I go into the forest a ways, to get out of the wind. I jump a herd of mule deer, and they pogo up the game trail ahead of me. I keep walking, up to the top of the ridge, to scan the alpine country for as far as I can see, looking for an upright figure coming down out of the mountains. He has probably found sign, I think, or maybe he's even watching a bear, and can't move without scaring it. I don't want to get too far from the car, in case I miss him.

An hour later, I go back down to the car, get in, and drive up and down the road for several miles. I know that Doug's feeling bad about being late. I can just feel it.

After an hour or so of cruising, I turn onto a dirt track that leads to a ridge with a view to an immense and wide-open alpine basin. Far below, I can see the great jungle of conifers, forested gorges, dolomite cliffs, waterfalls, more cliffs — all north- and west-facing, dark and impenetrable. Surely he's not there, I think, but somehow I can feel his presence in that jungle.

If he were up high, above the tree line, he could see the road from twenty miles away, and I would be able to spot him, moving across the short grass and rocks, from almost the same distance. I start back down the dirt track, into the dark trees. Yet another hour has passed and dusk is moving in, and it seems like a race to see which will get here first, darkness or the storm.

Just before dark, a lone truck, another woodcutter, passes me coming down from the Divide. He flags me, leans out his window, and informs me that my friend is out on the road about a mile or two back. It's beginning to spit hail, which bounces off the woodcutter's arm and off the truck's windshield, bounces off everything like popcorn. I thank the woodcutter and watch his battered blue truck pull away.

When I drive around a curve, I spy Peacock walking in the middle of the road. He is bedraggled, hailstones massing in his beard and thin hair. He looks to be in shock, staring round-eyed and gape-mouthed like a catatonic.

"Bad day," he says when he lifts the hatchback of the Subaru and tosses his bramble-torn knapsack into the back. "Bad, bad, very bad day." He says this as if he were scolding a dog. He's wet from the knees down, his boots are drenched, and he's shivering. "This has been a bad, very bad, not good day," he says, getting into the front seat.

We ride back to camp in silence, Peacock blinking and staring straight out the windshield at the swirling hail as if disbelieving that he's still alive.

"I've been lost before," Peacock says finally, "but never like this." He shakes his head vigorously, almost ferociously. *"Never."* He looks so rattled, this woodsman, that if he were to tell me that aliens had tried to abduct him, I think I would believe him.

The day started out well enough, Doug says. He had a bit of a hangover this morning, and still had an uneasy feeling from having gone into the lodge at Platoro last night, but as soon as he started walking, "burning off the poisons," he began to feel better. He crossed a creek and almost immediately found himself in good bear habitat. He worked his way into a stand of trees, picked up a promising game trail, and followed it along a rushing creek into deeper woods, watching the trail for tracks and scat the whole time. Around midmorning he stopped for a snack — a tangerine — and was sitting by the creek eating it when he looked down and saw a skull half buried in the moss and duff. He pulled it out of the decayed leaves. It was a bear skull.

It looked like a black bear's skull, Doug thought, but he couldn't be sure. Whatever it was, it was old. He put it in his day pack, finished his tangerine, and went on his way. About five minutes later it started getting colder and clouds moved in, spitting drops of rain. Not long after that, the woods turned strange. He got into the worst blowdown jungle he'd ever been in. He followed the creek down and around a corkscrew chute, crawling on his hands and knees, and before he knew what had happened, his mind was rubbed blank.

His sense of direction was gone. The sky, what he could see of it through the fierce jungle, was uniformly purple. There wasn't the faintest angle or glint of light at which to cast a guess of his bearings. Doug says there was moss growing everywhere, not just on the north side of trees and rocks. He kept falling and hitting his head on things. He slipped and sprained his ankle. He'd never been lost like this before, not knowing north from south, and he'd never been so uncomfortable, so frightened. The woods were spinning, coming in at him. The harder he pressed on, lunging against the many thick arms tangling and restraining him, the worse it got and the farther away he was carried, caught in the belly of the San Juans. He crawled and stumbled through the dense jungle for hours, until finally, exhausted, he lay down on his pack and tried to figure out what to do, where to go, all the while aggravated by the notion that I was confidently awaiting him.

Peacock knew he'd eventually get out. He would not be able to get out on time, and in fact might be hours or days late, but he *could* get out. He was totally uncertain about where his route would take him. If he continued forever, he'd end up on the east side of the mountains, near Kansas, or perhaps he'd find himself near Steamboat Springs, or even Salt Lake City, having taken a northwestward tack. What if he walked sixty or so miles in that other direction? Would he end up in Durango? Santa Fe?

Peacock says the thought came to him suddenly and clearly, like the lucid last thought of a drowning man: *Throw the skull away.*

Was it a grizzly or a black bear? It doesn't matter. Looking back from the safety of the Subaru, Peacock thinks that for the skull to have such power, the bear must have met a bad end. It died hard at the hands of someone evil, perhaps. Or it was poisoned, or gut shot, lingering in the woods, absorbing pain, gathering spirit before dying.

In the vortex of the jungle, Peacock stood up, took the skull out of his pack, and threw it over a gorge. Immediately, he says, he felt a peace returning. His senses kicked back in — smell, hearing, touch, and a sight beyond the constricted tunnel vision of the displaced — and the way out of the woods was revealed to him.

He crossed the creek, hiked up a ridge, and climbed a tree. From the tree he saw the clouds part, giving him a glimpse not only of the valley below — he'd gone way too far down the mountain — but also of the north slope to his left, the one he needed to cross and ascend in order to intersect our road. Still, he said, even with his reason back, he kept having troubling thoughts, or the troubling *echoes* of thoughts, like faint memories, of bears being wounded, chased by dogs, young bears being shot. He didn't linger.

His head began to split with the worst headache he'd ever had. At one point he considered lying down and dying, but he kept on going, and the farther he got from where he'd tossed the skull, the better he felt — and that's when I found him, right at dark, walking down the middle of the road with his zombie look. He wasn't sure what had been revealed to him along Skull Creek, only that it wasn't pretty, and that it was bigger than he or all of us.

I call it the past.

✳

Peacock and I reach camp after dark. Marty broke camp during the hailstorm, packed our tents and sleeping bags and hid the remains of the fire. He's run into a bit of a problem with his car, however: he can't quite fit the wheel back on.

Peacock opens a beer, grabs some tools, and crouches next to the bare axle. He holds a penlight between his teeth as he fools with the axle and hub, trying to pop some of the springs back in place. The hail has turned to light rain, and it feels like we'll never get out of here. The mystery of bears, their absence, seems to be pressing down — not just expelling us gently but booting us out, teaching us a lesson.

"A bad, bad, not very good day," Peacock manages to mutter while holding the flashlight in his teeth. He's sitting in a mud puddle. Thunder rolls in from the mountains on the other side of the valley. Lightning cracks against the cliffs a mile distant, bounces all around, and then another wave of thunder rolls in to meet us, and electricity raises the damp hair on our arms, on the back of our necks. Keetina, muddy as a catfish, leaps into the back of Marty's car, slithers across his backpack and up into the front seat, where she hides, dripping and panting. She jumps onto the steering wheel with both paws, blaring the horn.

Doug slides the wheel hub on, and it fits snugly. He bolts it in. We're moving like an Indy pit crew now, wheeling the tire over, lifting it up, bolting it on. At last we're firmly back in the twentieth century. We jump in our cars and drive out just as the deluge hits. I ride with Doug in the Subaru, and Marty and Keetina are in the Volkswagen in front of us.

We drive out of the mountains, and it is like being pulled through a car wash. I feel certain that the fury of the grizzly, all grizzlies, is upon us. The audacity we had, to think we could come sailing in here, walk faintly across a few of the bears' mountains, and discover a bear!

What we found was good habitat and one very big, blurred track.

Peacock talks as he drives, trying to keep Marty's taillights in sight. He has a cigar in his glove box, and we share it. Flashes of lightning illuminate the road ahead of us, an endless wave of washboard ruts.

It's as if the disorientation in the woods today, which temporarily rubbed Doug's mind blank, has opened up new loops and folds, caused old memories to surface. He tells me a story I've not heard before, about when he was a boy living in Michigan, and how he stalked the woods then, too. He was good at finding Indian burial places. When he was in his early teens, he was a celebrity in the Upper Peninsula because he'd figured out all the factors that told the Indians where to bury their dead — soil type, drainage patterns, exposure, and so on — intangibles of spirit as well. And when he mapped where he had found "his" burial places, it became evident to him where others would be. The University of Michigan newspaper heard about him — Doug kept finding the most amazing artifacts — and they came to the Saginaw Basin to find him, looked at his maps, and on the basis of his theory were able to get a big grant to hire graduate students to go out and prove that his maps and his theory were correct — which they were.

Another story, skipping ahead to the present. There's a counselor at the Veterans Administration hospital in Tucson who believes that Doug's alienation from the modern world was exacerbated by Vietnam. Based on several meetings with Doug, the counselor thinks he might be eligible for a partial disability, since he's got to get out and walk every day — across the desert, over the mountains, along the coast.

Some may call Doug one of the walking wounded; Doug prefers to call it "walking point." Edward Abbey walked point for the environmental movement, Doug says. It's a military term for the soldier who is at the front of his platoon, on the lookout for enemy activity. The point draws the first and heaviest enemy fire.

Over in Southeast Asia, you often couldn't see the enemy, or couldn't tell where they were firing from, once the attacks started.

Over here, back home, it must feel as if he's never gotten a chance to rest, as if he may never get a chance to rest. As if he's out front again, trying to help rescue the West and the wilderness from its present state of endangerment, its decline.

A battle wherever he turns. Roads, dams, mines, clear-cuts . . .

"All I have to do is go down there and sign a paper," he says, "and it's over like that — they'll start giving me money. I talked to my guy there and he says that's all I have to do."

Doug looks out the window at the storm. "I don't know," he says, and it startles me to realize he feels regret or loneliness about always being the outsider. He will probably not sign that paper until he has convinced himself that it's gone — peace — and that his landscape, for the remaining years of life he has left, will always be a civil war. Un-rest.

His nature is that of peace. It is only that he has been disturbed, and like the bear's, his fury and fight are directed not out of hate or war-like feelings, but aimed solely at reclaiming that peace, that territory which has been compromised, been stolen.

*

The rain has thinned by the time we stop at the paved junction to part company with Marty. We shake hands, hug, and then Marty's gone, off north to Boulder, on the other side of the San Juans. Doug and I turn south, to head back to Betty Feazel's. We still have one last duty: to file a report with Betty, who has been watching the mountains anxiously in our absence, looking up at them every day and wondering if there are still any grizzlies left.

The old house on the hill has a nighttime glow when we pull up the long gravel drive. I'm still thinking of things in terms of connections to bears, so it's no surprise that I feel like I'm returning to my den. We hurry through the rain, stamp the water from our boots, and step inside the bright warm home. Betty welcomes us with hugs, and Bruce and Lucy are there, too.

The old farmhouse has lots of windows. A view of the San Juans across the meadow is illuminated every time lightning flashes, the mountains looming right in our faces, seeming closer when glimpsed in those quick bursts of light. Doug remembers something, turns and goes out into the rain again as quickly as he entered.

He ducks back in with another burlap sack full of chili peppers. "It's bad luck to enter a house without bringing a present," he says, and the sweet and hot fumes of the peppers fill the small house instantly. Part Santa Claus and part trickster, Doug carries the sack to a pantry closet — it weighs fifty pounds, and I'm not sure Betty will be able to move it from that spot. We all go into the dining room, sit down at the table, and unroll maps, with the mountains leaping in and out of our vision with each flash, and thunder rattling the thick, time-warped windowpanes.

"Well?" says Betty. Doug takes his stocking cap off, rubs what's left of his wild hair, blinks and grins and then rubs his face. This is the kind of reentry he can deal with, being among friends.

After a long pause he says, "We saw a big track." The Baileys and Betty look at each other with as much delight as if we'd been *charged* by a bear. "I couldn't say yes but I couldn't say no, either," Doug says — meaning that he couldn't be sure whether it was the print of a grizzly or a huge black bear. Uncertainty of this type is rare for Doug. Betty again shows us the photo of the bear track taken in last year's snow. The paw print is nearly a foot long, and also suspicious, but the pictures are fuzzy, and details are hard to make out. Doug looks at them again, for about the hundredth time. "Quite frankly, what we saw is much better than these," he says. "These are good, these are maybes, but what we saw is good, too."

"Did you take a picture?" Bruce asks. "Can you go back in and make a cast?"

Doug shakes his head. "It was in the middle of a trail," he says. "Hunters would have seen it. We had to rub it out. We did take a picture." He turns toward the window and smiles as a bolt of lightning hits the meadow. Thunder rolls up the hill and rocks the old house, jiggles and tingles the blood in our arms. "What we saw is at least as good as what else has been seen. We just went in and out real quick, found that track and a couple of old digs. It's good country. We're just going to have to go back and spend a little more time — real low-impact but high-intensity searching. I'll put together some volunteers, teach them what to look for."

In *Wildlife in Peril,* John Murray describes the Colorado Division of Wildlife's findings of its two-year investigation in the early 1980s, most of it done on the ground, rather than from the air, because of a lack of funding. (Doug's biggest problem with that search, in addition to the fact that it was abandoned early, was that it relied on bait-trapped snare lines. These grizzlies, he says, would have learned, by the late fifties and early sixties, to avoid bait.) Murray writes:

The empirical results of the two-year investigation can be summarized as follows: 1) At one remote headwater location, in an open-timbered spruce-fir stringer between two avalanche chutes, at approximately 10,665 feet, a probable denning site of a grizzly bear was found. The interior was partially collapsed. The tunnel was approximately two feet wide and two feet high. The den showed remarkable similarities to grizzly bear dens in the northern Rockies. No bedding material or hair was found. It apparently was excavated and then abandoned; 2) Approximately fifty yards west of an alpine lake at 11,460 feet, a large feeding site for love root was found, pockmarked with many digs, and approximately a foot wide and half a foot deep. Old daybeds and root-filled scats were found in the vicinity of this site, which was located in an open-timbered sidehill park. Age of this probable grizzly bear excavation was estimated at three to five years; 3) Another dig, for biscuitroot, was found nearby on a xeric (deficient in moisture) slope at 11,653 feet. Biscuitroot has been identified as an important staple in the grizzly bear diet in the northern Rockies. Six digs of similar size to those described above were found. The age of this probable grizzly bear excavation was put at one to three years; 4) Another large dig was found about a half mile southeast of a major peak in the study area. A marmot had been dug from its den and several very large rocks were moved. The age of this probable grizzly dig was put at one to three years; and 5) in August of 1981, an outfitter reported an adult grizzly (blond, with dark legs, weight estimated over three hundred pounds) and two cubs in the headwaters of a major river in the study area. Personnel who visited the site found a large dig for osmorhiza.

*

Our maps lie outspread on the table, still smelling of woodsmoke. Doug runs his hand across the rumpled paper, showing Betty and Lucy and Bruce where we went, what looked good and what did not, what had been grazed ragged by cows and sheep and what still held hope. Doug tells the story of the bad day again, and Betty listens sympathetically, touches his shoulder, and then asks if she can get him something to drink. It turns out she has a variety of wines in her basement, and as she lists them for Doug, his eyes light up. He cocks his head and listens in the manner that I have seen him listening to birds in the forest.

"A merlot," he says, taking off his map-reading glasses. He places a stem of his glasses in his teeth. "I love a good merlot."

"I do too," says Betty.

Doug decides to stay the night. I have to leave right away for northern Montana to meet a friend, and so reluctantly I say goodbye. I long to stay, for the wine and the friendship, a hot shower and a bed, with the rain beating against the windows and snow up high, up where the grizzlies are.

Doug gives me a quick hug, as does Betty. I shake hands with the Baileys and step out into the rain. I drive all night through an electrical storm, torrential rain and wind. The storm has swept in from out of the mountains and across the high plains desert. I push into it, heading back west, back toward its source — on through its smashing, cleansing fury.

We found small clues on this trip, but the thing that tells me most strongly that there are grizzlies left in the San Juans is not paw prints, digs, or sightings of scat, but that yellow light in the farmhouse on the hillside, with a woman waiting inside for news.

PART II

›››››››››››››››››››››››››››

REVELATION

Wildness is a civilization other than our own.

— HENRY DAVID THOREAU

Throughout the following year I dreamed of grizzlies in the San Juans. In the winter I dreamed that they were sleeping. It was like sharing a secret with them, now that we believed they were still out there — and it was like holding a responsibility, too, believing that their future depended on how well we protected their country.

Doug and I know full well that there are armchair snipers, do-nothing naysayers who spend their entire lives bitching and bellyaching about how bad things are but who never get out and try to change or fix them. We know there will be a few whiners who say, "Aren't you hurting the bears and the mountains by drawing attention to them?" But none of us can effect change by being silent. We must turn the relentless flood tide of the way things are going, of the certain extinction that lies ahead for the bears if the system is not changed — if we don't protect, and reconnect, larger wilderness cores.

In the spring that followed, I dreamed that the grizzlies were awakening. Walking through the woods of northern Montana, I would imagine what they were doing in Colorado at different times of the day. They were leaving muddy tracks in the fast-melting snow; they were feeling the sunshine and the cold wind. I wondered if there were any new cubs — three new cubs, perhaps.

In the summer, when I went camping and traveling near my home and then on up into Alaska, I continued to think of those bears. A relationship was forming between us — I know this, though I don't know why. I understand now, in some indefinable way, why Indians preferred not to speak the bear's name, or the names of dead relatives.

While up in Alaska, carrying a hundred-pound pack over the Brooks Range, back and forth across the Continental Divide, I developed unusual symptoms in my left eye, and then in my hands and feet. I began to see electric blue bolts of light out of my left eye, flashes and short-circuit sparks of green and yellow, as if I were a robot and some of my wiring was coming loose. My feet and hands began to go numb, and I had significant memory losses. I also tended to mispronounce words, saying "saddle," for instance, when I meant "salad." When I left the mountains, I went to an eye specialist in Fairbanks who told me that it sounded like I had either a brain tumor or multiple sclerosis, and that I should see a doctor when I got back to Montana.

The checklist that ran through my mind went something like this, and not in any particular order: my mother was real sick, and I didn't want to cause any added grief to our family; I hadn't finished my first novel yet; I hadn't yet read *War and Peace* or *Go Down, Moses;* and my wife, Elizabeth, had just become pregnant. It was this last item that really sank in and permeated my thoughts.

I underwent an MRI scan, which failed to pick up a tumor. I went to the library and checked out books on MS. One of them said that a low-fat diet helps MS patients in some cases. I leapt into my new diet like a man swan-diving off a cliff. I gave up eggs, butter, meat, and cheese, and lost twenty-five pounds. Some days my eyes — both of them by this time — saw the electric field of blue sparks, while on other days I felt human, not like a machine.

In mid-September my eyes are still sending my brain the shorting-out messages. I continue to look at things as through the shadows of a ceiling fan that chops the light into shimmering pieces. It would be beautiful if I weren't so worried.

Is there a grizzly in the San Juans, hiding out in the big country, like a story you might have heard your grandfather tell when you were young?

Last year at this time, before my first trip into the San Juans, Elizabeth went to the doctor with an ache in her side. X-rays revealed a shadow, a cyst, a blemish — a mystery. We were going to have to wait a couple of months, then get another x-ray. This was a shadow I had carried with me into the San Juans, but a shadow that evaporated into yet another of life's blessings. It turned out to be nothing.

The heart of the story is this: Are there grizzlies — even one — in the San Juans? But part of the story too is the people who are trying to help decide this question, and the way they've fallen in love with the country there. What started out as a straight-line approach — to find clues, even proof — has become something wilder, something fuller. Whereas in the beginning we were all just roaming the woods looking for tracks and scat, what is instead developing, like deep-growing muscle, are friendships.

The mountains have always been here, and in them, the bears. To leave out the part about humans in this story would be to reduce the San Juans to a lesser wildness. And in my seeing new friends, I am seeing at least as much of the grizzly as if I were to crawl over a hillside and peer into a meadow and see one out there, grazing.

So Elizabeth got better. It was just a shadow. And now, this September, she's three months pregnant. My mother has leukemia, a bad kind, the kind you don't get better from. Last year she had it too, though I did not know it then. I've been down to Texas every month, visiting her.

These immense moments. Already, those great snowy mountains have become a yardstick for my life.

One good friendship in life is more than many of us ever come away with. I have been blessed with a few, and have been given another, Peacock's.

I am going to make a lot of friends in the San Juans. It is going to be like standing under an apple tree and shaking it. Fruit is going to come tumbling down.

Bears? What bears?

Don't coast is one of the things I've learned, looking for the San Juan grizzly. *It's all being stolen from you.* Don't blink, don't look away, don't be slothful. Keep your eyes open. Look closely everywhere. Pay attention!

The longer we can go without finding a bear, the more I am going to learn.

*

I fly to Albuquerque, where an old college friend of Doug's, Forty-Mile Ray, is going to meet me. From Albuquerque, Ray and I will drive to Pagosa and intercept Doug on his southward drift.

Ray and Doug last saw each other twenty-five years ago, when they were college seniors in Michigan. They were roommates, and in fact spent their next-to-last summer in the San Juans, mapping outcrops at a geology field camp. It was one of those usual college friendships — beer drinking, bullshitting, loyalties, allegiances, young men learning the world together and running as a pack before they all fell away into their new lives.

A quarter of a century! I admire Ray for taking the initiative to get back in touch with Doug, to keep that connection. Ray had read about Doug in an airline magazine; Doug's book, *Grizzly Years,* had just come out. I admire Doug for not giving up, for inviting Forty-Mile Ray out to Colorado to go into the woods with us, to camp and look around and perhaps find something.

Doug explained how Ray got the name Forty-Mile: that's how big his family's farm in Ohio was, forty square miles. Corn and wheat, and beef and dairy cattle, all run by Ray's family, with hired help. Then the farm was sold, piece by piece — taxes, foreclosures, the whole Midwest story. Only Ray's brother remains in farming now, on a small fraction of the original farm.

How important are our lives, anyway? How important is one remnant species? I believe that, to the eye of God and to the spirit of those mountains, man is nearly indistinguishable from bear, and that it is more than metaphor to say that we may as well be looking for ourselves.

As the universe cools, as humans move in to fill the place of the bear, we become the bear.

And how important are our lives? Beneath the Colorado stars, up at eleven or twelve thousand feet, each of us only one in a population of four billion, how important are we? Can we live in such a way as to keep the universe from cooling, or to keep the power and grace of a mountain bear from being lost to us? Do we have any more import than a mushroom or a spruce cone? The fact that we can even ask that question means the answer is yes, but we must not coast. We must hurry and look, pay attention, learn, if we are to earn a glimpse of the mystery of above and beyond.

We must see something that, up to now, we have not been seeing.

It's good to see Ray in the airport. He's wearing a tam-o'-shanter, one of those tartan plaid hats. He looks as tame as Doug is not. A quarter century ago, Ray ran track, but that was another man, another time. Most of Doug's friends are musclebound, strapping, but Ray looks like a . . . well, it doesn't matter what. He's friendly.

On the drive up to Colorado, Ray tells me about the last time he saw Doug — how Doug got in his jeep and headed west, out of Michigan and into his new life, while Ray stayed behind and stepped into the world of, shall we say, commerce and responsibility.

"I wish I'd gotten out here," Ray says as we drive north into the mountains. "I got married and had kids and never quite made it out."

Doug got married and had kids too, I think, but I know what Ray's saying. The "out" is what he's talking about, the crossing over, which, though better late than never, is easiest if done while young, when the body is fresh and strong.

✳

We drive to what was Betty Feazel's place, the At Last Ranch. The long gravel road winds through hay fields and past black, burnt-out lumber that marks where Betty's house stood for over fifty years. Not long after we left last year, her old farmhouse burned to the ground.

Her daughter and son-in-law, Lucy and Bruce Baizel (they've changed their last name), are building her a new house, a dream house. They busied themselves with plans for it the day after the fire, but what was lost can't be bought back, and the harder they try to assuage Betty's grief, it seems, the gloomier she gets. She's living in an apartment in town, and we will not see her on this trip.

Bruce has converted his office, a tiny cabin on the hill, into Citizens' Search, the headquarters of the group he's created based on Doug's work last year, and of which we are now members. The group's goal is to find data — proof — that might be missed by government agencies. With postcards, feathers, rabbit fur, and newspaper cartoons taped to the walls, the little cabin has the feel of a sixties hippie place, a place of youthful hope and energy. But with all the topo maps pinned up too, and all the reference books, it also has a military aspect to it, though not in a threatening way.

Also joining us this year is another of Doug's old friends, Jimmy Stearnberg, a TV-news guy, a maker of documentaries, who's been asking Doug for some footage of this story. And there's a new friend of Doug's, the grizzly biologist Dennis Sizemore, and one of Dennis's friends, a computer genius named George Fischer. Dennis and George are coming down from Salt Lake City.

Dennis, Doug tells me, is burned out on the heavy-handed approach of wildlife biology, the excessive reliance on radio collars and live trapping. Dennis was a biologist for the Border Grizzly Project, in northern Montana, and he's done his fair share of fieldwork. He's tired of the politics of wildlife "management," tired of the various agencies' lack of respect for the spirits of the animals they're supposed to be taking care of.

Earlier in the year, Dennis had gone down to Doug's house in Tucson with an antelope tenderloin and a bottle of George Dickel. He'd knocked on Doug's door and asked Doug if he'd help Dennis start a conservation school to be called Round River, based on Aldo Leopold's land ethic, discussed in *A Sand County Almanac.*

Dennis felt that the San Juans would be a good place for students to begin work, studying not just the presence (or absence) of grizzlies, but the whole ecosystem. A place where future land managers and biologists could learn to develop a sense of ethics along with science.

Doug ate the antelope steaks and drank half the whiskey and said that he wasn't "much of a schoolmarm," but that he'd try and help out.

I have no idea whether I'll become friends with Dennis and George; if anything, I'm a little annoyed at how crowded the project is becoming.

This is because I have not been paying attention.

M exican food and a dark bar. We're waiting for the rest of our party to arrive, having a margarita or two. Michigan's playing Notre Dame, and we watch for a while as Michigan surges. Doug waves at the screen, pleased with Michigan's lead, but promises loudly to all in the bar, and with a perverse pride, that Michigan will "choke, choke, choke, that's what they always do." He makes a gagging sound, and people dismiss him as a know-nothing crackpot, I can tell — until a few plays later, when Michigan fumbles a punt and Notre Dame takes it in for the go-ahead score.

Later in the day, we are finally all assembled, more like quail than hunters. We camp beneath the giant ponderosa pines behind Bruce and Lucy's cabin, watching the stars through the tops of the old forest, listening to the creek and the wind. Already it feels as if we are getting closer to the bear. Six people is too many to go into the woods together, so we'll split into two groups of three, although I don't think we'll find a grizzly by being stealthy or wily. I think the moment will come when the bear decides for it to come.

We will be looking for tracks and for scat, but that is the intellectual part of the trip. The heart of it is awareness that the bear could still exist — a thing that is almost as important as the fact itself.

Someone with more New Age lunacy, mooncalf sensibility, someone more head over heels in love with nature than I, might argue that this awareness is what the story's about — not the presence of four or five hundred pounds of flesh and blood in the assemblage known as a grizzly bear. That this story is about spirit, not flesh. Still, I'd like to find the physical part, the flesh part.

"Those who have packed far up into grizzly country," writes John Murray, "know that the presence of even one grizzly on the land elevates the mountains, deepens the canyons, chills the winds, brightens the stars, darkens the forest, and quickens the pulse of all who enter it. They know that when a bear dies, something sacred in every living thing interconnected with that realm . . . also dies."

Something sacred. I love my family, and beyond them, my friends. I love the wild, the big wild, the grizzlies and wolves. What I am going to have to learn is to love *everything,* all the way down to the smallest participant and the most abstract stranger. I must fall deeper in love with the act of breathing, and the honor of being alive.

I guess I am already a bit of a mooncalf after all. And running with this crowd, among these snowy mountains, is not likely to improve upon the situation — is not likely to convert me, say, to the sensibilities of a young Republican.

*

Two weeks after we left the San Juans last year, having found our one big track, the one that ran perpendicular to all the black bear tracks, a ranch hand named Dennis Schutz saw in the San Juans what he believes was a family of grizzly bears. Schutz has lived and worked all his life in these mountains and has seen all manner of black bears. The bears he saw were different, Schutz says. Those bears were grizzlies.

Schutz was on horseback — elk season would be starting soon, and he didn't want anyone hunting the woods before the season — a sort of self-appointed guardian of the forest. Schutz says he had sat down on a rock at the edge of a clearing to eat when he saw three large bears come into the clearing downslope of him, about a hundred yards off. Schutz watched them play in the meadow. He says they had high humped shoulders and round, dishlike faces.

The bears played for about twenty minutes. He thought it unusual for such large, full-grown bears to be acting like cubs. Then the mother bear came out — also hump-shouldered and grizzly-faced — and disciplined the "cubs," herding them back into the forest. "She looked as big as a horse," he said.

In describing the episode, the Durango writer Dave Peterson reported, "Schutz returned the following week with a Colorado Division of Wildlife Official [Game Warden Glen Eyre]. Even though a herd of elk had since passed through, trampling snow and soil, the men documented meadow digs and several bear tracks. One hind footprint was clear enough to measure: nine inches long by five inches wide. An average adult grizzly rear track is about nine by five and one-half inches."

The area around this meadow, this mountain park, is where we will investigate, almost a year later. Believing strongly that Peacock's theory is correct, that any grizzlies left in the San Juans are slightly different from other populations — that they remain in family units longer, that they are shier — we will camp a couple of miles away from where Schutz saw the bears, and sneak in. We'll be as quiet as we can be, and won't eat or drink in the area. We'll spend only a couple of hours a day looking for scat and then we'll head out well before dark. We'll try to be no more obtrusive to the bear family than a sparrow that pauses on a branch, then moves on. They'll smell us coming in — a grizzly can smell an elk carcass at a distance of eight miles or more — but they'll smell us leaving, too.

Is it possible for a grizzly family to rely heavily — through generations, even centuries — on a certain drainage as a primary home? Dennis Sizemore and Doug, despite their scientific training, remain open to mystery, to possibility. Doug says that it could very well be the case. If it is, our chances of finding something by returning to the same spot where Dennis Schutz saw the bears, and the bear sign, are as good as they are going to get. Much better than last year, when we went sailing off into the high country with no real direction, just one old, cold rumor, and were nonetheless able to find that one big track.

*

We camp down low that night, truck camping, miles from our destination. We'll take day packs up into the mountains each day and keep our base camp low. As we're setting up our tents, a herd of cow elk runs through camp, galloping through the aspen and waist-high grass. Although it has snowed up high, it's been raining down in the valley, thunderstorming. We're camped below a sheer cliff; we'll have to find some cleft tomorrow if we are to work our way into the high country. The maps do not show that any such passage exists.

The woods are drenched, and steam rises from the trees as we fix supper. Doug's nervous that we're so *exposed,* even after four-wheeling in several miles up an old logging road. He has his summer cache of firearms with him, pistols that people have given him. He passes them out to us for protection, in case any preseason elk hunters pass through, angry at our intrusion into "their" territory. While at first this may seem like excessive paranoia, I've learned to attend to Doug's hair-trigger senses; when a man who fought with the Green Berets for two years says the woods have a strange feel to them, I am not going to tell him he does not know what he's talking about.

I'd prefer not to fool with one of Peacock's pistols, but he's so intent and cautious for his band, his clan, that I take the pistol more for his peace of mind than mine. It is a nine-millimeter Beretta with a full clip, brand-new, as fancy as a race car; Peacock keeps it locked in a carrying case, much like an executive's briefcase. He shows me how it works.

Because my tent is farthest downslope, I'd be the first to encounter trouble in the form of drunks shooting at our tents.

That night at the cooking fire — a little propane hisser on the tailgate of Doug's truck — I note that Doug is in that transitional stage between city nervousness and mountain euphoria. He and Jimmy Stearnberg, the filmmaker, talk over old times, though he insists that Jimmy turn off "the damn camera — I'm too tired to fool with it tonight, I don't want to look at it, turn that shit off." He reminisces with Ray, too, asking after an old colleague of theirs, Tim Benson, a.k.a. Needle-Dick Benson.

Peacock tells of how, after he arrived out west and hooked up with Edward Abbey, Abbey heard Peacock speak of Benson and became amused by, even enamored of, that unusual nickname, despite the fact that Abbey and Benson had never met. Once, Doug was at a book signing of Abbey's — perhaps it was at a store back east — and Doug was sitting at the table with Abbey. There was a long line of people waiting to get their books signed, and suddenly who should appear in front of them but Benson himself. He and Doug hugged, and Doug introduced him to Abbey as "Tim Benson." Abbey smiled but looked puzzled, says Doug, but then the light flashed and Abbey gave his great grin and shook Benson's hand more heartily than ever, and cried out in a big booming voice that stopped everyone in the bookstore, "Oh, *Needle-Dick* Benson!"

It is not a campfire if there are not stories.

Later, I lie down in my tent with my grandfather's down vest wadded under my head for a pillow. He had died four months ago, and the vest was the only thing of his I really had any use for that meant much to me; he always wore it when we went hunting together. It still has his scent, very faint, and I fall asleep thinking of how old he was, and of how young I am compared to him. What is it like to have another fifty-five years of life ahead of me?

My mother is fifty-six. I tremble to think that she may not have another year — may not have another four full seasons. Her life hinges on neutrophil counts, blast cells, lymphocytes, and platelets — the all-important platelets. Marrow infections, temperature readings. The stars, seen through my tent flap, try to give me solace, but I am frightened. I understand that existence must be defined by the presence of its opposite — absence and loss — but why do I see loss almost everywhere I turn? Who's in charge here, order or chaos? One bear, one Colorado grizzly, would kick this century of loss, of greed and of taking, square in its ass.

That night a storm moves into our camp. Bolts of lightning get trapped in the crevices above us, hurling sizzling blue light through the woods, not unlike the flashes of light in my vision. Peering out of the tent as the sky cracks open, I see another elk herd running through the aspen. I don't know which is worse, the thunder or the lightning. The lightning strikes all around — the nearest strikes less than a hundred yards away, so that the ground we're sleeping on is electrified, tingling, buzzing — but the thunder is bad too, so forceful it knocks me to my stomach. When I sit back up it hits again, closer, and again the roar of it flattens me. I'm pressed against the ground as flat as I can be, shivering as the heavens open. The storm is so lunatic bright, so ravingly close, it seems personal. More lightning strikes, six or seven in a row, each one accompanied by an explosion that deafens me, fills my mind with light, and I'm shaking. The earth has gone to war.

I wonder what's going on in Doug's mind, with Vietnam so far away and yet never gone, and I wonder what's going on with Sizemore's friend, Big George Fischer, whose greatest fear is being struck by lightning. I don't actually *ponder* these things — they're just synapse thoughts, as each smash of thunder and lightning presses me flatter to the ground. My only true thought, again and again, is the belief that I am at the edge of death and am about to pass through those gates. The hair on my head, face, arms, and back is standing straight up. My teeth will not stop their burning feeling.

Then the wind comes. At the time, I don't know it is the wind. I am disoriented — deafened by the continuous thunder, electrified by the charged air, and dizzy from the furious light show — but when the wind comes, it is more than wind: it is a solid pressure. It lands on my tent, steps on my ankle, falls across my back, and then begins slugging me, again and again. I am certain that territorial elk hunters or anti-grizzly people or drunken yahoos have waited for this storm to move in before attacking. Whatever is jumping up and down on my tent is landing so hard that it's bruising me.

The Beretta, I tell myself, with the fury of being attacked. I feel for and find the briefcase, open it in the shuttering lightning-boom of daylight, shove the clip in. When the next attack comes, I jam the pistol up into what I perceive to be the assailant's kidneys, jam it hard and shout, "I've got a gun!" I tighten my finger on the trigger, certain that whoever it is, is going to start shooting at my tent.

But there is no more rib kicking, no more leaping on my wet, sagging tent. I have scared the marauder back to the woods. The tingling in my teeth subsides and the wind passes on. My body hair begins to settle, and in the steady calming downpour that follows, I understand that it has only been the wind, only nature, and I am a foolish man. Nonetheless, I feel cleansed, and I lay my head down on my grandfather's vest and sleep deeply.

*

We look like wet rats the next morning, blinking in the bright sunlight and holding our tin cups of steaming coffee with both hands, grateful to be alive. We all agree that it was the most violent thunderstorm any of us has ever encountered. The woods around us seem bruised, aching from the storm, but the sky is pure and blue, and the mountains shine with a new cleanliness. What some may perceive to be pagan, superstitious stargazing — finding signs and omens in all of nature's moments — may also be called simply good common sense, the heightened or focused alertness of the hunter. Doug looks up at the cliffs, smiles, and says, "Well, that's just right. The mountains wanted to get us ready to see the bear, wanted to get our attention."

Great cobbled spires of ancient conglomerates and metamorphosed breccia tower into the sky along the cliff's upper ridges. These spires are more resistant to erosion, more enduring than the surrounding rocks.

Deep, rich, dark timber — fir and spruce — grows wherever the mountain has dissolved to make soil. But those magical spires remain, like giant spirits turned to stone, watching us, or guarding the bears.

"The Hoodoos," Doug says. "The Hoodoo Mountains."

Dennis, who used to play linebacker in college, turns his calm eyes toward the cliff and seems to appraise heaven itself with his quiet look. Despite his height and size, his face looks like that of a priest. George also has that curious look, as we all watch the cliffs and think about what Doug just said.

Forty-Mile Ray and Jimmy Stearnberg do not seem to be evincing this reverence, and I, the writer, always playing it both ways, too rarely making the leap, just watch and listen, like a predator in the cliffs.

In fiction, you want to get as close to your story as you can, but in nonfiction, in real life, you're better served if you doubt everything, revere almost nothing, and always protect your heart. It's a fine thing to say, *Yes, yes, the mountains were preparing us before opening their secrets to us, for revealing the bear,* but the wary journalist, the careful predator, can't go pursuing every twig crack in the woods, every surge of the heart. If everything's sacred, then the world itself is sacred. To a journalist, however, it is in a story's best interest if only certain moments are sacred.

I love the spirit of bears and the spirit of the wild, but I also have taught myself to hang back and watch. I don't think too much of the moment when Doug and Dennis and George give the mountain their silent prayer. Yet I feel that my resistance to the sacred is slipping. It must mean we are closer to the bear than I think.

"Chemicals!" Doug cries when the reverie is over, sloshing more coffee into his cup. We hear the hiss of the propane stove perched on the tailgate. "Chemicals!"

We mill around a bit, patch our battered tents, and pack our knapsacks for the long day ahead. We'll move slowly through the woods, looking at the ground and trying to be as ghost-like as we can. We hope the elk will have a game trail we can find, some secret series of switchbacks. Dennis Schutz, the ranch hand, rode in from the other side, but we're afraid we'll blow the bears from their cover if we go in that way, up the creek. Instead, we'll go in straight over the top. We will take two different routes around and up over the cliffs — Doug, Ray, and Jimmy in one group, and Dennis, George, and I in the other. I notice with the hesitancy of a man walking on ice that my shutter vision is better this morning, and wonder if the storm's terror pounded it out of me.

Dennis Sizemore was a bear biologist for the Blackfeet, east of Glacier National Park. Doug has told me this and a few other things about Dennis, saving me a lot of questioning; to say that Dennis is quiet or soft-spoken is an understatement. He just doesn't talk, period, if you're a stranger. And if someone else is talking, he never, ever butts in. If a whole group of people are talking, he watches each one and listens. I've read that Indian tribal councils were and are run the same way, with each warrior being assured a time to speak, and it seems that Dennis is comfortable and familiar with this form of courtesy. (It could also be that he thinks I'm a white asshole who has nothing to say, and is wondering why Doug has brought me into the mountains.)

George Fischer is also quiet around me — the social leprosy, the stigma, of my being a journalist — but I have been briefed that he studied science at the University of Utah, got a doctorate in physics, and is now working as a computer programmer. It turns out he's got an incredible aptitude for computers, and he's been made head of some big Salt Lake City company that designs and sells computer programs. George is thirty-four. He's six foot six, two hundred and sixty-five pounds, and is an earnest karate student (Dennis, too, is a black belt). George has a poet's heart but can be rough as a cob. He has a week's worth of stubble beard, this trip. I couldn't have imagined a more unlikely person to be a programmer.

We work our way quietly along the base of the cliffs, all but tiptoeing up the steady grade of game trails. So much meat, but so few fangs in these mountains. Just above an elk wallow, Dennis finds the first scat. It's small for a bear, but it's definitely bear shit, old and dried-out and black. We sit cross-legged in the sun-dappled forest and examine it, picking it apart to see what the bear's been eating. It's all leaves, grass, pine nuts, and ants. "These are the first pine nuts we've seen," Dennis says. He pulls one from the dry mass and hands it to George.

"This is how I saw my first grizzly in the wild," Dennis says. "I was up on the north fork of the Flathead, picking through a scat I'd just found. I happened to look up and saw the bear sitting on the hill above me, just watching me. He looked down at me all puzzled, like, 'What are you *doing?*'"

Dennis breaks the scat down further, as deliberate in that act as he is with his sentences. What he's looking for is hair, from the bear's grooming itself, then passing the hair along with everything else. Unfortunately, the U.S. Fish and Wildlife Service and the state of Colorado do not yet consider hair analysis as legitimate proof that grizzlies exist. But we'll accept it, because it is the truth.

Our job, therefore, is simple. We don't need to kill or catch any bears. We just have to move through millions of acres, pick up all the bear shit, find the hairs in that shit, and then send them to a lab for analysis.

If there were another way to do it, we would. But it's almost impossible to find a track — too many rocks, too much vegetation — and we won't try to catch them with bait.

Even if we could entice a bear with bait, Peacock and Sizemore wouldn't stand for it. To encourage the San Juan grizzlies to come in to bait, even for presumably "good" reasons, would be to unravel their previous century's learning, and would probably doom them.

We — Citizens' Search — have been trying to proceed with humility. We fear that our efforts may somehow embarrass the official agencies that ended their searches. It's not a macho thing, but we're worried that it could become that — a contest to find proof, a prideful thing, which would hurt the bear. To the agencies' credit, the longer and harder we search, the more open they seem to be to the notion that we might find something.

Schutz, too, is one of the wild cards. I believe Schutz's story, and I think that Glen Eyre, the warden who went back with him and saw the tracks, believes him too.

"This is the part that my wife hates to see me doing," Dennis says, breaking the scat open. He passes bits of the dry dung to George and me when he's done with them.

"No hair," Dennis says, "just grass and leaves, a few ants, and those pine nuts. But no hair. Not a very grooming-conscious bear."

"Must've been a slob," says George.

Dennis hands him the last piece. It smells like dirt. The ant corpses in it glitter like jewels. George puts the samples in a zip-lock bag, as he did during the past summer, when he was up here for thirty days. He'd number and date the samples and send them to Dennis in Salt Lake, with map coordinates of where they were found. Dennis would pick them apart, collect any hair, and send it to the pathologist in Wyoming. All this stuff traveling back and forth through the mail.

"Seal that tight," Dennis says. "That last batch you sent was pretty fermented by the time I got it."

"Sorry about that," says George. "It rained for a month. Never had time to dry it."

"My mailman doesn't like me anymore," Dennis says. "You can smell it even in the package."

"What would you say is the diameter of the tubes?" George asks, making notes on the plastic bag.

"Beg pardon?"

"The *tubes.* How thick?" George says. "Tubes" is his own nomen-clature.

"Oh." Dennis gets it. "About an inch, or three-quarters inch."

We leave some of the scat behind, to return to the soil and to release its chemicals, to avoid erasing that animal's sign.

It begins to rain. What's it like to eat ants, I wonder.

Beginning in the 1930s, the Texas folklorist J. Frank Dobie collected oral histories about grizzlies in the Southwest, and some of the infor-mation he gathered told of how trappers would kill a bear in the spring, just after it had come out of hibernation, and how they'd find live ants in the bear's stomach. The ants were the first thing the bear must have eaten, ripping up a log to get at them. The bear's stomach would not have begun producing acid yet, so the ants would still be alive, crawling around in the empty stomach.

Dennis points out claw marks on aspen trees; on one aspen, a bear climbed almost all the way to the top. He says that sometimes bears climb trees to get away from other bears, but they also climb them for fun, and for the view; younger bears do more play-climbing than adults. The aspen is at the edge of a bluff, looking out at the whole valley below. It's the sort of tree any of us would have climbed when we were young.

We pick our way up a steep game trail, hoping the elk have found a pass into the cliffs that NASA's satellite topo-mapping system missed. We jump a herd of elk only fifty yards away, beautiful butter-yellow cows and mahogany bulls and calves. Everywhere we turn, it seems, we are bumping into elk.

We need a wolf. The earth is being trampled, flattened, by hoofed things.

Have I said that I hunt? Surely I've mentioned it. But this is embarrassing, this many elk. There's no challenge, no sense of accomplishment, if you "get" one. It becomes just shooting, not hunting. When you are the only predator in the woods, you are another step — a big step — closer to the act that so many hunters claim to revile, the so-called immorality of going to the grocery store and buying your meat cellophane-wrapped. Hiring someone else to do the killing.

What's dangerous about the grocery store is what's dangerous about the San Juans: there are few, if any, other big predators present. None against which to measure yourself, none to learn from. To be like a wolf, a bear, or a lion — even if only for a few days a year.

Absolute power corrupts absolutely; hunters, more than anyone, need other predators. There is room for both, room for all. I will not go on and on about the diseases of overpopulation or the reduced size and quality of the animals. I will just ask hunters and outfitters to consider this: the best thing about hunting is the hunting, not the killing.

You must squeeze the trigger and clean and pack out and eat your prey for the act of hunting to be complete, but the best part nonetheless remains the way you enter the system when you are hunting, the way you are totally within it. The crack of a twig, the hammering of your heart, the soft brush of hoofs through new snow, the pause, and then the big head with the antlers appearing through the brush like a crown prince, an emissary that has come down to earth to allow you to hunt and eat and kill. *"Let the predator love his prey,"* writes Jim Harrison.

A blue grouse hoots on the game trail ahead of us. It flushes and then two, three, four others also rise through the trees, sounding like feathered dynamite. Jays follow us briefly, wondering why we're in their woods, wondering why we are moving so cautiously and slowly.

We find a small tight pass between two of the eerie rock spires — a tiny grassy saddle, smaller than a typical back yard — and we lunch there. My feet are numb again, and the vision in my left eye is once more like looking through a spinning fan blade. Raisins, pretzels, apples, bananas. I pour the fat-free food down me like gasoline through a funnel, hoping, if it's MS, to burn the scar tissue off the nerve endings. It's sobering and daunting: if the grizzly bear is on the ropes, what chance do frail humans have of making it through this world?

We descend the back side of the saddle, through the sweet-scented firs and across the steep talus above Heartwood Creek. Dennis points out dainty bobcat prints in the damp gravel below the game trail, paws so round and small and perfect that they're as magical as seeing the cat itself. We look around, trying to see the view as the bobcat must have seen it just hours earlier.

Across the steep gulch of Heartwood Creek, wild roses hold late-season blooms. The sun is warm, driving steam off the creek and up through the rock spires, wisps of fog rising. Across the creek the trail gets even steeper, and the rich evergreen color of the fir trees invites the mind to consider the lush richness of rotting logs. Shade, and secrecy.

It is George who spies the skull. It rests on a promontory, high above the whitewater rush of the creek, on the west side of the gulch. The skull is positioned so that you have to be standing in this one spot to see it; from anywhere else, trees or rocks block the view. Heartwood Creek, like so many of the San Juans' drainages, corkscrews down and around. Only at this bend of the narrow gulch, and only by looking up through this particular combination of rocks, this cleft, are you able to spot the skull. We sit on a rotting fir log and look at it with our binoculars, then move up and down our steep slope trying to get a better perspective.

It looks like the skull of a huge bear, shining white beneath the blue sky. It rests on a flat spot in the shade of a few scraggly wind-busted spruce saplings. It is nestled in a little low-walled natural turret, the kind of outcrop you see in westerns, where the bad guy ambushes the stagecoach as it passes below. Above the promontory, the hillside has slumped, a seventy-degree slope of fine scree. Above that, the fir trees begin again.

We trade binoculars and speculate for at least half an hour. We can see the eye sockets, can see the rounded cranium. It's too far away to see the individual sutures of the skull, and we search as hard as we can for a way to say that it is just a white rock. But it is not.

We cast about for bones around the skull, bones that would be strewn by wind and time, by birds and coyotes, but there aren't any. The view is such that it seems the skull has been placed where it now rests. It has the look of the sacred to it. And the view: as if standing guard over all that lies below.

We try to be reasonable, scientific, even dismissive. We try not to believe. It's the size of a horse's or cow elk's skull, except there's no muzzle. The roundness of the cranium indicates it could be a human, but it's far too big. It can only belong to a large bear. The eye sockets and the round sweeps and curves that only skulls possess — this, in a country of angularity. We stare and marvel at it and try to convince ourselves it is not what it looks like.

"A golden eagle could have dragged it there," Dennis says. It would be a day out of our way, even if we could figure out how to get over there. We'd have to go around to the top of the mountain, then rappel down, blind, hoping to stumble onto the promontory by hanging over the cliff and dropping down with ropes.

"But I don't think it's a skull," Dennis says, almost angrily. He seems irritated at the mystery, the nearness and the distance, the unproveableness. I think it's definitely the work of humans, because of the way the skull has been placed, its loneliness. For me it has a sinister feel, a message of warning.

Our job is to explore this lower country, to try and scout a way up, and then to meet with Peacock and the others in midafternoon. Feeling out of sorts about the skull, we resume the search for scat. Dennis tells us to look for day beds, too — sometimes bears will leave fur behind where they nap. "They like flat spots on trails, usually beneath a big tree," he says, "so they can see and hear and smell anything coming from either direction."

We're walking as quietly as we can, creeping, but every now and then we crack a twig or pop a little limb. A boot slips on a rock. Just one of those sounds is all it takes to move the bears, and any other wild animal, out and away.

Dennis points out squawberries, with their chrysanthemum-like leaves — a preferred bear food. "Lots of nice little nooks and crannies here," he says. "Good female bear country." A mother bear could raise her cubs without ever having to expose herself. Up to eighty percent of all the grizzly bears killed each year by humans in Montana are females, due to the fact that the demands of raising the cubs make the mothers more vulnerable, more visible. "Nice little pockets," Dennis says, looking across at the steep forest and the intermittent gray cliffs. He's spotted two possible routes to try, though there won't be time to attempt them today.

We loop down and around the largest spire, and at its bottom we find Doug and Ray and Jimmy. It's hot and we're drenched with sweat. We show Doug the scat, and we tell him about the skull. He's attentive. "It doesn't have to be recent. It could have been put there by Indians," Doug says. "This is great country. Magic. Maybe an old bear went out there to lie down and die. *Great* country." He and Dennis compare notes on the diversity of the area, all the different grizzly foods they saw.

We climb up through a chute, looking for another saddle that will take us back to camp. We make two rest stops on the way in, both in the shade. In one small grassy park, there are more claw marks on an aspen where another bear has climbed to look out at the valley. "Bears like a good view," Doug says. "It's true. I can't tell you how many times I've found a nice big bear shit right at an overlook."

At the second rest stop, a peregrine falcon plunges out of the sun at us, ripping straight down and across the sheer cliff face. Peacock holds his hands out, cautioning us to freeze, and still the falcon dives, coming close enough for us to hear the rip of feathers and the slash of its being before it pulls out of its swoop and hurls away. Peacock turns to us and is glowing, hunched over and grinning like a kid with a secret. He places his hand over his chest, takes his baseball cap off, mops the sweat from his big balding head.

It's the second peregrine that's bombed me in a month. The first was in late August, back in the Brooks Range, up above the Arctic Circle. I had just gone over a steep pass and the falcon came right at me, veering away when it was only a few feet from hitting me.

"These are really special mountains," Peacock says, still clutching his heart. "I can feel a lot of things going on here."

Dutifully, I make notes in my little spiral memo book. It occurs to me that in trying to inventory and measure magic, all of the luster is lost. The rip of the falcon's wings is muted, the dive-bombing of the moment becomes stale. Perhaps I should throw down pen and paper and run hard into the woods like one of the bears — ripping and clawing at a hollow log and consuming ants and grubs and the rich mulch of the earth. In abandoning the story, I might end up seeing more, and learning more.

You would think that the writer, the predator, would have a way to see everything, to catch all movement. But now it strikes me, in the presence of Peacock and Big George and Dennis, that perhaps the way of prey, not predator, is the way to see the most — with the wide eyes and heart of a deer, rather than the tunnel vision of the wolf, the bear, the fox, the journalist.

Steam rises from our wet boots as the day grows warm. Doug and his crew have found a route too. Tomorrow or the next day, we'll head up to the grizzly meadows, again in two bands.

*

We're cooking communally, preparing the eternal tsamba. The stars are out, and the war of the lightning bolts the night before seems ancient, like something from years ago. Campfire stories are being told in low voices and whispers beneath the bear's mountains. We're marveling at all the elk we've been seeing. With the mating season going full bore, the bulls are bugling. Dennis tells about how the wildlife science textbooks have a term for the bulls who don't bugle, but who instead run in and try to steal cows from a herd bull while the herd bull is out trumpeting his lust. "Sneaky fuckers," Dennis says, is the term the textbooks use for those bulls who come in silently, without announcing their presence. "I was a little startled to see that term used in a textbook."

Dennis is positively garrulous tonight. "I'm moving off from George," he says. George's tent is pitched right next to Dennis's. "He'd snore three times, then fart twice. It was a symphony. There were elk circling our camp all night long, investigating his mysterious noises."

We're all sharing a bottle of George Dickel, and Peacock's leaning back in his folding river chair, his prune-wrinkled, squid-colored feet propped up by the fire. *The woods.*

"Those are the ugliest feet I've ever seen," George says, to shift the talk away from the indelicacy of his night noises.

Peacock looks genuinely hurt. "They've just been marinating for a couple of days," he says, a little defensively. "When they're dry, they're nice feet."

"My ice is thirsty," Dennis says, rattling the cubes in his cup.

George, who's Bogarting the bottle, hunched over it like it's a cached elk carcass, straightens as if from a reverie, says "Oh!" and starts the bottle back on its rounds.

Of course some of the talk is about bears. So far, George is the only Citizens' Search volunteer who's seen a bear. At the time, George and another volunteer, Martin, were a couple of weeks into their search, not far from where we're camped. George had been trying all day to persuade Martin to drop down off the ridge they were walking, as it looked like a storm might come in — a lightning storm, George's nemesis. George's father is a survival school instructor for the army, and ever since George was a boy, his father has cautioned him about being hit by lightning. Martin, though, was glorying in the static charge up high, the smell of approaching rain, the tingling ionization of flesh and air, the nitrate smell above the tree line.

Finally George talked Martin into getting off the ridge. As they hiked down through ten thousand feet, George saw it. A light rain began to fall. The bear was big and brown and had a round face, an odd light-colored face like a ghost, he says. The bear looked at them for two or three seconds before slipping away into the bushes. It was getting dark, and because they were excited about seeing the bear, they decided to make camp right there. They saw the bear again the next morning, coming up the game trail toward them. It gave them a three-second look, of which all George remembers are the impressions "big" and "brown." Then the bear veered away. George and Martin broke camp and veered away too. Despite the rain, they hadn't been able to find any tracks.

We listen with reverence as George relates his story. Doug's nodding, thinking about George's description of the bear's round ghost-colored face.

"It could have been dirt-faced from digging," Doug says. "When Mexican grizzlies got down to just two individuals, they were still digging. It's a hard instinct to override. It sure lets you know where they are." We are reminded that the state of Colorado's search turned up some digs following the Wiseman bear's slaying.

Peacock explains to Ray and Jimmy, who have been a bit out of the loop of the wild lately, about the bear's importance to the culture of man: how our ancestors probably learned all kinds of survival skills from watching bears — how to store food, for example, and which grass, nuts, berries, and mushrooms to eat. Then, Doug says, "Someone killed a bear," and man fell from grace, has been falling ever since.

We finish Dennis's bottle of Dickel and are pleasantly surprised to hear that George has another bottle in his pack. Jimmy goes over to the pack, sifts through its contents. We hear the clink of glass. Jimmy grunts, lifting George's big pack, putting various items back in place.

"What've you got in here?" Jimmy asks.

"Stuff," says George and shrugs. "You never know when you're going to get your ass snowed in," he says, his father's son.

Doug starts pulling his socks and boots back on now that the fire's dwindling, and George informs him that one of his boots has caught fire.

"I burnt zee boot!" Doug cries in a low mournful voice, leaping up and stamping it out. These boots were a gift from a French mountaineering firm that wanted Peacock to advertise its product. "Zee boot," he keeps murmuring, "I burnt zee boot!" We pass the Dickel around and settle in, our lives crossing and growing together. We're not endangered or threatened, or don't feel it, anyway, even though there are only six of us, which may be how many grizzlies there are left.

More bear lore. Back when Dennis was the biologist for the Blackfeet, up on the eastern boundary of Glacier National Park, he met one of the tribe's elders, a medicine man. On his first day of work, Dennis went to this elder and explained what he wanted to do: trap and radio-collar the bears so their movements could be followed. (This was before Dennis became disillusioned with conventional wildlife management.) The elder refused permission at first, but after a year of Dennis's giftings to the elder of elk, deer, and antelope, permission was given to trap and collar the bears. First, though, the elder gave Dennis instructions to follow — a ceremony to perform each time he caught one of the bears.

"Basically," says Dennis, "what the ceremony did was tell the trapped grizzly that the elder had nothing to do with this, and to apologize to the bear." Dennis performed the ceremony every time.

*

It's late, and we're tired from climbing along the steep ravines. Doug goes off to his tent to read by flashlight before going to sleep. The fire now is a small circle of glowing coals. Dennis tells one more story, about the first time he met Doug. It was in Montana, on the west side of Glacier, at the Northern Lights Saloon. Dennis was in there one night having a few drinks — this would have been not too many years after the Vietnam War. Peacock ran into the bar wearing his backpack. Doug bought two twelve-packs of beer and went over and asked Dennis who the hell he was, it being Dennis's first time in the Northern Lights.

"I'm with the Border Grizzly Proj—"

Peacock jabbed a thick finger at Dennis. "Leave the fucking bears alone," Peacock said, then shouldered his backpack and picked up his beer and ran the twelve miles through the woods to his fire lookout up on the mountain.

"Back then, he ran everywhere," Dennis says, meaning not like a jogger but like an animal — a deer, a fox, a coyote, a wolf. "I never saw him when he wasn't running. Even if he just had to cross the street to go to the grocery store, he'd run."

*

We're up early with the elk. We can see them moving through the trees in the thick gray fog like ghosts, as if we've descended into the underworld, where huge armies of elk move on forever. We start up the trail in silence, changing our group a bit: Ray and I hike with Doug, while George and Jimmy go with Dennis. The ends of my feet are numb again, and my eye is seeing flashes of green light.

For some reason — the logic of a witch doctor, perhaps — I have told no one of my strange ailment, only of the diet. The thought has crossed my mind that I will really throw a wrench into everyone else's plans if I drop dead, like a head-shot buck, at ten or eleven thousand feet.

We climb quietly, then stop at a rest spot. Doug finds himself talking about the town of Moab, Utah. For thirty years the town has been a sort of spiritual way station for him as he passes from the desert to the mountains and back. He lived in Moab for a while and had some of the best times of his life there with Edward Abbey.

The valley below (were it night, we'd be able to see a light or two, the everywhere reach of man) reminds us of how frail and vulnerable all of the West is. Moab, which is in the middle of the desert, is especially doomed, a Phoenix-like target. Doug and I view the new McDonald's there — with its billboard, "Come visit Moab's other arches" — the way a doctor would look at the first blemish on an x-ray of a smoker's lungs. Doug informs us glumly that there is now a mountain-bike trail through the Nature Conservancy's wetlands holding on the Colorado, one of the last wetland areas along that great river. And there's a Hilton, Doug adds. Moab's identity once stemmed from its hard uranium-bust dirtiness. But all that has begun to change, and there are no geographical barriers to inhibit the town's growth. I'm reminded of Catherine the Great's silly dictum: "That which is not growing begins to rot"; perhaps it was one of her descendants who dreamed up the Moab billboard slogan.

Resting in the autumn light we come to an old but ignored consensus: only the West's eternal wrestle for and with water stands to serve as any kind of inhibition to growth. Even now, developers are casting eyes on the Yukon, wanting to ream it straight down to the American Southwest.

What if Pagosa Springs becomes similarly exploited? What would happen if outsiders come in and buy and change and take control of that small town?

Such dire thoughts lift us to our feet and get us going up the mountain again, side-hilling through the dense timber and climbing, always climbing, through and over lovely rotting fir trees, out of which grow vibrant ferns. It's cool in the jungle, but our Moab discussion has brought civilization in too close to us. Like the bears, perhaps, we feel the urge to flee, to put more distance between us and there, and the only way to do that now is to go to the top of a mountain, and farther into one of the last forests.

We'll spend the day combing the trees below the mountaintop — and well below the grizzly meadow — looking for more sign. It's hard work. We go through water a quart at a time, passing our canteens around and then stopping to refill them, with the aid of Peacock's elephant-piss pump, at each stream crossing.

The Hoodoos tower above us, serene and wise in the high winds. "A magic spot," Peacock says again. He spreads the map out on the ground, pulls his reading glasses from his shirt pocket. His blunt hammered hands remind me of hoofs. The fingernails are flattened from rock-climbing mishaps, convex instead of concave. Sweat drips from his forehead, splatters the map at the place he is trying to study. He curses, takes his glasses off, and mops his face with the crook of his arm.

We'll work our way back past Heartwood Creek so we can point out to Doug the skull on that eroding, inaccessible lookout. One of the things I love about Doug is how, if you catch him at the right time, he will listen — not just to the woods or to the wind, but to you. In those moments, a word or thought or observation from the most naive member of a group might carry the most interest for him. He probes all aspects and angles for mystery.

When I tell him that I think it is a bear skull, put there by some local as a taunt or a warning to those who would come into the area looking for sign — for rumors have come from this area ever since the death of the bear in 1979 — when I tell Doug this imagined nonsense, he cocks his head and does not blink, just stares at a point on the ground as intently as if he is listening for the approach of enemy soldiers. "Could be," he says, after he has run it through his mind. "Could be."

We climb different game trails, switching back to the north, whereas the day before we'd gone south. We examine new parts of the same woods, knowing always that we could have missed a scat or a print, missed it by a few feet or gone right past it as we gazed at one of the spires.

The differences between Doug and Dennis, the two bear men: Dennis moves haltingly, pausing often to listen and look, like a hunter — a cat, perhaps. Doug moves quickly, more eagerly, erratically, almost as if he's at play in the woods — deciding that in some soft-path stretches it is okay to lunge down the trail, while slowing to a creep in others, and noticing, pointing out, commenting on everything.

The bears must be here. The habitat is still here, and in the presence of Doug and Dennis, the spirit of the bears is here. How can the bear *not* be here?

How can the world end?

Speaking of dead things, of men and women and civilizations gone to dust, yesterday Doug found a tiny chalcedony arrowhead, almost snow white but translucent, nearly invisible when held up to the sun. He put the arrowhead in his pocket to take home for his daughter, Laurel, certain that, because of its shape, it had not been used for hunting — chalcedony's too soft — but rather for ritual purposes. All day long, he says, he could feel its magic.

But then he lost it. There was a hole in his pocket, and the arrowhead, with its sharpness, must have found that spot and cut its way out, returning to the San Juans — "as it should," Doug admitted, though he had very much wanted his daughter to have it. He's cheerful about the loss, good-natured. The loss of the arrowhead confirmed its power.

It's comforting to think that there was someone long ago in these mountains giving care and attention, passion, to such a small thing.

And if Doug is not killed by some great-toothed predator while out in the woods — when Doug leaves this world, what to do with him? He carries a little card in his billfold, like the ones used to alert physicians to certain medical conditions. It is the size of a credit card, and in fine print it says:

FEED THE BEARS

I, _____, being of sound mind but dead body, do hereby bequeath my mortal remains to feed the Grizzly Bears of North America. Respect my body. Do not embalm! (A little mustard would be appreciated.) My family and friends have been instructed in how to deal with my corpse. They may be reached at: phone (___)___-___; address _____.

Please put me in a deep freezer if I must be held for a few days. Should my family refuse to claim me, or should I be indigent at the time of my demise, please explain to the County that I can be mailed to a wilderness (as evidenced by the presence of grizzlies and/or wolves) for a lot cheaper than I can be buried in a pauper's grave.

Please remove my eyes, kidneys and heart for use by the living, but retain my liver because I think Griz would like that the most. I love you all. See you in the Spring!

*

When we reach Heartwood Creek, we show Doug the bobcat prints, which are now barely distinguishable. We travel a bit farther down the trail, excited to show him the view of the skull. When we get to the bend in the trail where we can most clearly see the skull and the lookout point, we see the skull is gone. Everything else is the same: the turret arrangement of weathered rocks, the gravel (no fresh earth slump has swept it away), the scraggly brush around the turret.

"Maybe a bird carried it away," Doug says. He's kind, recognizing the hugeness of our loss — that empty falling feeling. He shrugs. "Who knows about these things? That's just part of the system. Like my arrowhead yesterday — it just wasn't meant to be carried home." He smiles gently; because we are crestfallen, we don't get it. "It just wasn't meant to be carried home," he says again.

"It wasn't a patch of snow or ice," I say. "It was a skull."

"No, I believe you, I *believe* it was a skull," says Doug.

"It wasn't an elk's skull, and it wasn't a horse's," I tell him. "It was either a bear's or a human's. And it was big — bigger than any human skull I ever saw."

Doug smiles, lifts his hands. I did not get the message then, though I get it now: mystery is as real and necessary a component of these mountains as are the tangible elements of bears, rock, sun, trees, water, ferns.

*

In the dark woods, we move more and more like hunters on the verge of discovery. Always right at the edge of that change, crossing over from mystery into knowledge.

"Look," says Doug, holding out one hand to freeze us. He points to the ground.

"I don't see it," I tell him, searching for a bear's big footprint among the ferns and moss.

Peacock looks at me incredulously. Ray crouches and squints, staring at the moss, also trying to make out the pattern of a print.

"Look!" Peacock says again, so vehement that I think spittle may fly from his lips. "You don't see them, do you?" he says.

"Them?"

Peacock bends down hurriedly, takes out his pocketknife, and begins cutting at the base of something below the moss. My eyes focus and I see the fresh-cut shock of white plant flesh, plant meat. From out of the moss he lifts a beautiful fluted bright orange mushroom, holding it in the palm of his hand as a preacher might hold a Bible.

"Chanterelles," he hisses.

I'm red-green color-blind, which is common in males and left over from a time when visual sensitivity to tone, rather than color, was important — the hunter's eyes picking out the slight variation in winter grays: the deer in the woods, the mottled ptarmigan against the snow and rocks. I can see red clearly and green clearly, but I can't pick out red or bright orange when it's against a field of green. In Peacock's hand, however, the chanterelle leaps to life, glows in the shape of a wine glass filled with sunlit wine.

We pull out our plastic bags, heretofore reserved for turds, and move along the steep mossy slope beneath the towering trees, plucking the chanterelles where we find them. Thoughts of bears vanish as the biological imperative of the hunter-gatherer stirs within us — hydroxyl groups binding to blood-rich red atoms of iron, for all I know, providing a giddy, heady, pulsing wave of search-discover, search-discover. We are oblivious of everything but the ground. Clouds may have passed across the sun, the wind may have changed direction four times, bears may have wandered through the woods not thirty yards away, and we wouldn't have noticed, down on our hands and knees.

Some of the chanterelles are only tiny buttons, and others are lovely spreading fans, clusters of them as large as two hands. By crawling, examining each patch of ground, breaking it down into an area with borders about ten inches square, I'm able to pick out a few of the fungi.

Peacock's an expert on mushrooms — he's been a plenary speaker at the National Mushroom Festival — but neither he nor anyone else really knows what factors conspire to create mushrooms in a certain place at a certain time.

You can map where oil and gas will be found, two miles underground and a hundred million years ago, and you can also look at a topographic map and tell where certain species will be found — deer, bear, ponderosa pines, aspen. But no one can really plot mushrooms on a map. You have to go into the woods to find them. Of course they need shade and moisture — it's been a wet summer — but they need odd combinations of dryness before the moisture, a certain angle of light, certain temperature variations, to unlock the biochemical movements and secrets of the entire forest substrate. You need luck, is what it boils down to; more mystery.

For hours, we move around on our hands and knees, upslope and downslope, all around the shaded ridge we're exploring, crawling and moving our heads from side to side, and digging. Surely, if anyone were to see us from a distance, he would think that our dark shapes were those of bears.

*

A little higher, then, when we can find no more mushrooms. We climb skyward, still staying below the grizzly meadow to keep from spooking the bears. It feels good to hike the steep slope through the trees, climbing straight up rather than using switchbacks, pulses hammering — an exquisite kind of smoldering through which we will emerge, breaking into the other side of nature, the animals' side.

C ampfire. No bear scat was collected by either party today, but we covered a little more ground, combed a little more ground. A light snow would help, but it was rain that fell today, even up high. There aren't many light snows high in the San Juans, just heavy ones, which, once started, don't stop until May.

Are we here for the bear, or for each other? It's both, and it's pleasing the way this shifts and weaves. We've filled our day packs with chanterelles, and now Doug empties our cache onto the hood of the truck. The orange heap is moist and fragrant, and we scoop up the mushrooms and let them trickle through our hands like coins. Jimmy Stearnberg gets out his video camera to document the men, if not the bears, and Doug's smile fades immediately. Even the unpracticed eye can see Doug bristling, can see the nostrils flare, the eyes widen.

Doug, courteous at first, ducks out of the picture, steps off into the darkness of the trees and busies himself with some make-work, examining the guy lines of someone's tent. In bears, Doug has told me, this is called displacement behavior, and it's done when the bear is nervous. He defines it as "behavior entirely inappropriate to its stimuli." Doug has filmed examples of this in grizzlies in the wild. In one sequence he filmed in Glacier National Park, a young grizzly walked right up to Doug as he lay in the tall grass. The bear came too close — so close that its entire face filled the camera frame, blurring the focus. The bear was suddenly as nervous as Doug, and it lowered its head, looked left and right — pretending the camera wasn't there — and began nosing around in the grass, pretending to look for beetles and ants, when the last thing on its mind was food.

Jimmy films the pile of chanterelles for a moment, then swings the camera around and heads straight for Doug and begins filming Doug's fiddling with the guy lines.

Doug grunts at this invasion and stalks off toward the campfire, to pretend to have a word with George, who is seated like a giant stone statue before the fire, a druid, content simply to breathe in the smoke and watch the flames. When Jimmy hurries after Doug once more, it becomes too much and Doug turns around and explains, tersely but politely — or what is polite for Doug, though it might be called brusque by someone else — that "tonight's not a good night for the camera."

Jimmy's still not getting the point, arguing that this is exactly what he wants, that this is great footage, this campfire stuff. I'm reminded of how last year, when Marty Ring and I went into the woods with Doug, I had to take my notes out of Doug's sight, to keep from spooking him.

This fear of the camera might be related to Doug's notion of living life instead of writing about it. You turn to face the camera, you proffer a goofy smile and mumble a few words — but in the meantime the current has passed by, and your soul is following it, while you're standing flat-footed before the camera, suddenly empty, thinking about all manner of things.

Doug explains all this to Jimmy in two very short and profane sentences, and Jimmy finally gets the point, or if he doesn't, at least he turns the camera off. (Later that night, Jimmy will stand Dennis and me before the lens and ask, "And what did you see in the woods today? Did you see any bears?" And like fifth graders during show-and-tell, we'll recite the day's events, a kind of enumeration that seems, afterward, to threaten the magic. *No, we didn't see any bears. We saw a bobcat print. We found some mushrooms. The sky was blue. It was a real nice day.*

*

We've got a little butter, a little canned milk, garlic, onion, pepper and salt, and a pretty good cache of mushrooms. Doug makes a dish that would cause a revolution in New York. We pass the pot around, fill our tin cups, tear off chunks of stale French bread Doug found in the back of his truck, and we cannot stop eating.

I remember the fried chicken I had in Costa Rica earlier in the year, a chicken that an hour earlier had been chasing grasshoppers. I remember trout I've caught in high mountain lakes in Montana — how I take some home in the long evenings, and how, only an hour or two later, I'm eating those fish. The same with huckleberries picked from the mountains, the same with elk backstrap . . .

We lick the pot when all is gone. The taste lingers so exquisitely, so deliciously, that for a long time we sit around the fire in a kind of stupor, not even bothering to open the cap of the Dickel, though finally, when the fire fades and the coyotes call, Dennis motions for the bottle, opens it, takes a swallow, sighs happily, and says, "Another day, another dollar not made."

That night, mushroom dreams. And meat to accompany the mushrooms on their journey through my body and into my cells. Dreams of steaks and seafood, and then more carnivorous images: sharks. Then a strange stew made by native peoples — puppy stew, a horrible thing to imagine. Dream images of Fairbanks, and the tundra, and then of Tucson and the desert — the extremes of the West, I realize in the morning, with the San Juans somewhere in the middle.

Mushroom dreams! I dream the old man in my novel died. If I eat enough of Doug's chanterelle stew, will my dreams match those of the sleeping grizzly in these same mountains? What if we, grizzly and man, eat mushrooms from the same hillside? Will the dreams come up from the earth, and go through us, in roughly the same manner?

*

The hiking and the mountain climbing are easy. It's the paperwork that requires the true strength. And Dennis, with his fledgling concept of a Round River program, a Round River school — Dennis, the old linebacker — will be doing the pushing. Pushing those masses of paper across the table, dumping them in the politicians' laps, and writing grants, assembling momentum, changing the very being of inertia. He'll schedule meetings with Fish and Wildlife personnel, with Game and Fish officials, with the Forest Service, establish that network, let them know we're here to stay. The wearier you grow, the harder you push, until you emerge with victory, which is the only catharsis. All the rest of us have to do is walk across the mountains. That's easy. It's the long nights up late in the office, the kind of work that lies ahead for Dennis, which will take all the strength he has.

A bit of that kind of work, in fact, lies ahead of us today — people work and the world of paper, rather than mountains.

It's not as if we haven't been out among people before. It's easy to paint us as woods savages, naives amid civilization, but it must be remembered that Dennis lives in a city of one million people. I grew up in a city of two million, and worked for a long time — long enough — in industry, as a geologist for an independent oil and gas company. And Doug was in that war. So we know about people. And today, rather than going back into the mountains, we've got to meet with some. Doug and Dennis are looking forward to one of the two meetings, but the idea of the other one is tearing them up, and they need to get it over with for reasons that have more to do with cleansing than anything to do with the San Juans, or the last few grizzlies.

I'm supposed to go with them as the voice of reason, and pull them off their antagonist if they get into a fight. Doug's gotten all wound up about this meeting, cursing and raging about the usual injustices, building a case against the world in the way that only Doug can, working himself into what Jimmy Buffet calls "a tumultuous uproar." I know ahead of time that no one will be able to bear the brunt of his increasing fury, and I find myself feeling sorry for this fellow whom Doug is going to confront in the day's ugly meeting.

Doug and Dennis decide to get the nastiness over with first. There's this environmentalist fundraiser who had promised Citizens' Search some expense money. The guy had raised about fifteen hundred dollars, which would go for gas and stamps and the hikers' food as they roam the mountains for weeks at a time. It's such a small amount; one thinks of the U.S. Fish and Wildlife Service's six- and seven-figure budgets, of their meetings in Denver, where a single plane ticket can cost that much.

That's the bedrock beauty of Citizens' Search and of Dennis's Round River concept. To say that the group operates on a shoestring would be hyperbole. It is more accurate to say that Citizens' Search operates on grit and the eyes of potatoes and fifteen-cent packages of Top Ramen.

The fifteen hundred dollars was promised, and half delivered. Half has been held back, pending — what? Successful completion of the project? Does the fundraiser — whom Doug, in his fury, keeps calling Suicide Ned — think that the volunteers will pocket the money and head to Mexico?

What Doug fears is that there are strings attached, that his presence will be required at some goofy wine-and-cheese function. Doug's afraid that Suicide Ned has compromised himself and told the donors, "Sure, my man Doug, the barking seal, will come up and get the money. He loves to sing and dance. You can meet with him, get your pictures taken with him as you present the second half of the check." In fact, Ned had mentioned something along these lines.

"I don't want it," Doug keeps saying on the drive out to meet with Suicide Ned. "I don't want it, I don't want it, I don't want it." He continues the previous night's diatribe, mourning the loss of purity and honor, assailing that part of human nature that always makes the slick intermediary try to get the most for his money, to play the angles. "I told him I didn't want any strings attached," Doug says. "Fuck the money! I told him about the project and that there were good people working long hours for free because they believed in it. If there was someone out there who wanted to support what we're doing, then fine, but there would be no song and dance. I don't have time for that kind of bullshit. We need to be up in the mountains, looking for the bear. If they can't understand that, then fuck 'em."

Doug looks out the open window at the summer hay meadows. There are elk down here too, grazing, even in midmorning. The sweet mountain scents of warm grass drift across to us. Dennis is driving, all the while watching the road and listening. Doug's twisting and jerking in his seat as if it's constricting him.

I find myself feeling sorry for Suicide Ned, who may be a decent, even a wonderful guy. Raising money for environmental causes is better than, say, raising funds for neo-Nazis, but Doug has no interest in such distinctions. The money was promised, and the promise was not kept.

We're to meet Ned at a ranch, one of the largest remaining tracts of private land in the San Juans. The woman who owns it, Lavinia, is keen on learning what she can do to help our cause. She is young and immensely wealthy, and about eight and three-quarter months pregnant with her first child; after meeting with us, she will fly to New York to have her baby. She's also invited Dennis Schutz — the man who saw the grizzly family — to join us.

We drive through forest and down along the river bottom — more meadows. We keep heading south and pass cattle yards, flat basins with grapes-of-wrath dust, woolly Herefords standing runny-nosed beneath the hot sun with nothing left for them to graze. The Herefords stare into infinity, waiting for the sound of the hay truck. A lean coyote walks through their midst, but they pay him no mind. The coyote walks up to one old knock-kneed bull, who must weigh close to a ton, and the two creatures stare at each other, almost touching noses.

We pull off to the side of the road and watch while cars and trucks whiz past, bound for Santa Fe. Then the coyote understands he is being studied, is hunted, if only with our eyes, and he turns his attention from the bull to us. We're parked about four hundred yards away. He sizes us up for half a second before whirling and running off in the opposite direction, running with that beautiful floating canter that coyotes have when they know they're on top of the situation but are still in a hurry. He kicks up occasional clouds of dust as he looks back, left and then right, never stopping, running until he is the color of the dry dead earth itself, and still running on back to the green trees and the cool mountains.

If I think that this momentary image of the wild is going to sidetrack Doug, I'm mistaken. The coyote is like a shot of freshness, but as we turn onto a gravel road and follow it through the heart of cattle country, Doug resumes his verbal attack.

After a dizzying number of turns, which we navigate by topo map, we finally come to the gate that leads to Lavinia's ranch. We drive for what seems like half an hour before we see her house on the hill, overlooking a river. The meadow grasses are high and green, and even Doug relaxes a bit when we stop for a moment on a wooden bridge to listen to the blue water rushing beneath us. Because Lavinia doesn't allow hunting on her ranch, there are a lot of elk, hundreds of them. She could use a few predators such as bears or wolves, what the academics call "megafauna." The dark hearts. This is a country that believes in checks and balances, right?

There's a huge lodge below Lavinia's house, a kind of bunkhouse operation with a restaurant-sized kitchen and a cavernous banquet hall for entertaining visitors. That's where we're to meet Suicide Ned, with whom Doug and Dennis have spoken only on the phone.

But we'll know him when we see him; he'll be the only one around. As kind as he is brutal, Doug wanted to save Ned from being embarrassed in front of Lavinia or anyone else, so he asked Ned to meet us at the bunkhouse down in the trees, beneath the giant cottonwoods.

*

It's like a detective movie, or an old western in which the whole town is deserted and the outlaws are stalking each other. We go from room to room in the huge lodge, beneath towering rafters, looking everywhere for the man Doug's been calling Suicide Ned for so long that in our minds it is now his true name. For some reason, we're reluctant to call out for him in the big-chambered lodge, so instead we split up and move in three different directions: around the perimeter, in the south half of the lodge, and the north half. But no luck.

We reconvene in the banquet hall, and because there's nothing else to do, we go into the kitchen to see if any food has been left behind. The water and power have been shut off in preparation for the coming winter, but maybe there's something we can borrow for the campfire: a clove of garlic, a can of crabmeat, a jalapeño pepper, a loaf of bread. We just finish snuffling through the cabinets when we hear footsteps down the hall. We scurry back into the ballroom and await Ned's approach.

Ned is tall, slender, and dapper with short black hair and an inquisitive face, a man who you might think would have sufficient moxie to quit his job as a lawyer or labor union spokesman or whatever, and hie off to Tibet on a spiritual quest. He's wearing dress shoes that go *click click click* as he walks across the ballroom toward us, and in his body movements I can read that he is surprised, maybe a little alarmed, that there are three of us.

He introduces himself with the grace of a diplomat. It is his job and his custom to move in monied places, as if among the very coins and bills themselves, and those are places that make our knees quake. He's got an intelligence about him, a brooding pensiveness, and also an alertness, like a gazelle or an antelope. We trade names and handshakes, and after that's done, Doug says, "Look, Ned, we need to talk. I'm not happy about the way you've been holding the money you said we could have."

Ned gets a puzzled look on his face. He motions to some chairs over by a window and says, "I'm sorry to hear about it. Let's sit down and talk about it." We all walk silently toward the waiting chairs in the far corner of the banquet hall.

Doug and Dennis and I sit with our backs to the wall. Ned sits in front of us, and right away Doug starts talking about how hard the volunteers have been working and how it's all been out of their pockets, and in the rainiest summer in years. Doug is working himself back into the tumultuous uproar that has been festering for the past eighteen hours. "Nothing's for sale here. I feel like you've really let these people down," he says. "That money was promised a long time ago, and you've been holding on to it for whatever reasons, and I'm sorry, we're just not going to be able to sing and dance for it, that's not why I'm out here. Where we need to be is up in the woods. I resent that you've hung us up this long, and I'm here to tell you we don't want your money. I'll find it somewhere else, somehow, or we'll just do without. We don't want the money." Doug makes a cutting motion with his hand. "We just don't want your money," he says, and then sits back and glares, red-faced, at Ned.

Ned is quiet for a good long time, long enough for the echo of Doug's words to have evaporated. Finally he responds. "Well, I can understand what you're saying, but I don't think you can speak for your friends, can you?" He turns to look at Dennis, who he knows needs the startup money desperately for Round River. Dennis looks right back at Suicide Ned, tightens his chest a bit, and says, "Yes, Ned, in this instance I think Doug *can* speak for me."

Ned's face puckers with surprise, but he recovers quickly. "I see," he says primly. "I have to make a phone call, can you excuse me?" He gets up and leaves the banquet hall.

Doug and Dennis have been presenting their sternest, most malevolent faces, but the moment Suicide Ned is out of the room they explode with laughter, clapping their hands over their mouths and making snorting, giggling sounds. Doug imitates Dennis's cigar-store-Indian grimace and mutters, "Well, in this case I think Doug *can* speak for me," and they snicker like schoolchildren who have, in a strange twist, sent their teacher to the principal's office.

All this fuss and furor over fifteen hundred dollars! I'm sure that Ned is accustomed to dealing with figures ten times, maybe even a hundred times, that amount. I'm sure, too, his worst suspicions of the woods savages, of the underground environmental activists, are confirmed: misanthropic, zealous, scruff-bearded yahoos with a latent capacity for violence. Perhaps Ned is thinking, as he dials his phone, that the wine-and-cheese party was not such a good idea after all, that although the donors might have enjoyed a touch of earthy idealism, perhaps Peacock and his strange friend would have given them an overdose.

Doug and Dennis are still giggling when they hear the businesslike *click click click* of Ned's shoes; they stop just as he reenters the banquet hall. Once more their faces become fierce and grim, like gargoyles. Suicide Ned says something conciliatory, like, "I'm sorry we couldn't work together," and goes on to explain how it wasn't really his fault; the donors had somehow done the delaying. He had been calling them, in fact, and leaving messages, but . . .

I believe him, but it's too late. Peacock has acted, and he feels better.

Perhaps it isn't Ned's fault, and maybe he's not such a bad guy. The important thing is, Doug has *acted,* and now he feels better. Strangely, I think Suicide Ned feels better too.

The four of us walk out of the lodge into sunlight. The wind is rattling the autumn-dry cottonwood leaves. These giant cottonwoods were planted when the ranch was homesteaded as a Spanish land grant, back when on any given day you could see a grizzly or a bison. Peacock looks relieved, almost expansive. He tilts his head to the sun, to the blue sky. "What a day," he says. "Is it great to be alive or what?"

We murmur our agreement. Suicide Ned splits off from us and strikes out across the field to Lavinia's house, where the second meeting will be held. We get in our car and take the long way around. Doug and Dennis are more somber now, realizing they're going to have to bring home bad news to the volunteers, but everyone feels better, immensely freer, and Doug says, "Lookit — *fuck it.* I'll get the money somehow. But we've just got to disassociate ourselves from that kind of thing." We know what he means, and we believe it, though it might sound holier than thou to articulate it. A fresh breeze blows in through the open car windows, a breeze that bends the top of the tall grass and calms, or so I imagine, every living creature over which it passes.

Well, almost every living creature. We enter Lavinia's compound — the back patio of her house — and find a group of people standing around talking to or being interrogated by what seems to be a madman in a camouflage outfit. Bristle-bearded, scrawny, and heavily armed, he's clutching an automatic weapon with one wiry arm and gesturing emphatically with the free arm. There is also a pistol and a knife on his hip, and another rifle as well as a bandolier of cartridges strapped across his back and chest. In the midst of this circle of people we recognize Ned and the obviously pregnant Lavinia.

It doesn't look like a life-or-death ruckus to me, just another strange scene in America. But Doug doesn't like it at all. As we approach the jabbering commando, he puts his hand out and motions us to go slow, says under his breath, "What's going on here?"

"Spread out," Doug tells us. "Don't make such a big target." We move in warily, unthreateningly, as if we'll try to talk the wild-eyed gunman into laying down his weapons.

Lavinia recognizes Doug from his pictures, and introduces herself and her ranch manager, Lily, a short, muscular, beautiful red-haired woman from San Antonio. Dennis Schutz, the man who saw the grizzlies, is there, as is what appears to be one of Lavinia's legal or financial advisers. And then there is Max, whom we will later call Machine Gun Max, and still later, Rabid Max.

"This is my neighbor," Lavinia says, pointing at Max. She walks a graceful tightrope between showing the minimum daily requirement of courtesy due a neighbor and evincing her amusement that a grown man is running around dressed like this. "Max has been up in the woods looking for trespassers. He says he saw some up on his mountain — he saw them through his scope." Does she mean a telescope or the scope attached to the rifle on his back?

"They were wearing brightly colored clothes," says Max, and he scuffs the gravel with his boot. "Hikers," he says contemptuously, and squints at our plain brush-brown clothes, our bear-hiding clothes.

"Wasn't us," I say. "We were a long way from here. Looking for grizzly bears." Lavinia rolls her eyes; I might as well have said, "Looking for the Antichrist."

"Max is worried it was poachers who'd come in to shoot elk before the season," she explains.

Max, who suddenly seems aware of how goofy he looks, all decked out for killing but with nothing to kill, summons a show of fierceness that will justify his presence. He clenches a fist and warns Lavinia, "You'd better watch out. They're up there. I'm going to go looking for them some more." He turns quickly, hops on his three-wheeler, and blasts off down the dusty drive, the roar of his engine echoing through the valley, bouncing off cliff walls, finally fading to nothing more than a whine. I think how lucky we are to be working where there may be only grizzlies, rather than commandos.

Then I think about the way weapons have appeared in this story. It was a bow and arrow that killed the grizzly in 1979. And then there was the pistolero in Chama, waving his piece at us as we investigated his junkyard one Sunday last fall. And now Rabid Max, a veritable walking munitions plant — though he, too, held his fire. How different it all might have been if the archer had held *his* in 1979. This, of course, is my belief — that that grizzly was shot before it charged and received the stab wound to the neck.

But we've been over all this before. It doesn't really matter who fired at what, or when or how. What matters now is that people like Rabid Max and poachers and regular fair-season hunters hold their fire. That with regard to bears, the guns remain unfired.

*

Once inside, Lavinia makes us feel immediately at home. There's a grand piano, a huge picture window, and walls and walls of books. Lavinia slips her shoes off and gathers her legs, lotus style, on the couch to listen to what we've found. There's beer in an ice chest, and, except for Lavinia, we all have one — that friendly sound of pop-tops being snapped. The purple backdrop of the Rocky Mountains makes us even more comfortable, and we begin to tell Lavinia our bear stories. We sip our beer and toast George and the others, who are back in camp or out on the trails searching for scat.

Dennis Schutz recounts his sighting for Lavinia, draws on a topo map for Doug, showing the boulder on which he was sitting when he saw the bears. Doug's excited, we all are, and what impresses me as I watch Doug is how intent he is, not on the checklist of the grizzlies' physical features — round face, short ears, long claws, shoulder hump, all of which Schutz saw — but on Schutz's *response* to what he saw. The fright, the electricity, the tension that Schutz felt — Schutz the cowboy, who's lived all his life in these mountains and seen hundreds of bears. Doug's not interested in fur color or snout length; any old black bear can have a blond coat. What Doug wants is the whole of the experience, the spirit of it.

"It was different, huh?" Doug asks.

"I knew they were different the second I saw them," says Schutz.

"And then when the big one came out — you could tell she was real nervous, being out in the open like that, and she was pissed at the other bears for being out there. They were big bears themselves but they minded her right away. And then they were gone. I didn't see them again."

"It wasn't until a few days later, when I was home, that I looked at a picture book of the bears up in Alaska, and Yellowstone. That's what these were," Schutz says.

"Great," Doug says softly, "great. It was something to see, huh?"

Encouraged by Doug's enthusiasm, Lily tells us about a dead elk she found earlier in the summer when she was fishing. The elk was a fresh kill, and something about the whole area made her hair stand on end. She was so terrified that she had to leave the place immediately. Lily's as tough as they come, we can see that; she's worked all her life on ranches. It's possible she's seen ten thousand dead animals. But only one, up above the tree line, had something about it that made her skin crawl, that told all of her senses unambiguously to leave.

Doug listens to this story with equal intensity, and I begin to understand that that's what he's listening to: not just the perceptions such as sight or sound, but to all of the senses, including our cells' memory, coded deep within us, from a time of cave bears, a time of dark nights and crackling campfires. The body remembers these gone-away things more quickly than the mind does in this, almost the twenty-first century, and Doug is listening to these people's bodies as much as he is to their words. I find myself listening harder, and it occurs to me that if you want to listen to people's bodies, then these two bodies, Schutz's and Lily's, are more calibrated to the wild and more reliable than most others.

Now Lavinia wants to know what our plans are: if we'll sue the Fish and Wildlife Service to recover grizzlies in Colorado, or if we'll even tell them of our findings. It's clear she's a woman of action. Doug and Dennis look at each other, shrug, and say they're taking it a day at a time. "We don't know what's out there," Doug says. "We just want to find out."

Lavinia pauses, reads the two men, understands in an instant that they don't have any bigger agenda or motive, that they — we — are simply drawn by the power of the mystery. A question has been asked of Doug — are there grizzlies still left in Colorado? — and Doug wants to see if he can answer it.

Cooperation with the agencies of authority — the blinking computers, the dripping coffee machines, the memos, the decisions, the meetings — all that will come later, and perhaps, if Dennis Sizemore's program is operating then, Round River and the community can participate. Perhaps Doug and Dennis can show everyone a hint of the spirit-force that lies behind the grizzly's rediscovery — the passion and care people still have for the land, a land administered to by people behind desks in Denver and in Washington, D.C. Doug calls these administrators "good people who started out in the woods, but who forget to get out anymore, people with good hearts who need to get back out and sleep on the ground."

Doug and Dennis can remind the authorities about the passion and power a mountain range, say, can hold for the human heart. They can remind the authorities that it's okay, more than okay, to release those wild stirrings. You don't have to repress them just because you're wearing a uniform. In fact it is wrong, perhaps even evil, to repress those feelings of wildness, of greatness, peace, and exhilaration.

We want to rediscover the bear in Colorado, even if we never find one. The notion, when you look up at the snowy peaks, that there could be one just over the next ridge . . . that discovery, or rediscovery, may be as important as the sight of humped shoulders and long claws, the bear itself.

All this Lavinia sees in a glance, reads that we are dreamers. Still, she's excited: she wants to *act*. She's asking about the reintroduction of grizzlies, wants to know how that would be done. Dennis explains the difference between hard releases and soft releases. The most successful reintroductions occur when the animal is given a soft release — put in a large enclosure in the woods with a minimum of human interaction until it becomes accustomed to its surroundings, to the scents, sounds, and rhythms of the land — the sound of that area's bird calls and the pull of gravity itself. An animal replaced in this manner is far less likely to bolt and strike out for home, crossing highways and villages in its panicked flight. Most successful of all is when a pregnant female can be kept in such an enclosure, can give birth in the enclosure, and even raise her young there for a few months. Nothing bonds a creature to a place like giving birth there. It's what defines "home" in the wild: the place where you make a stand.

"Would my ranch be a good place for that kind of thing?" Lavinia asks. Doug and Dennis tell her that the ranch's habitat is excellent, though we all can't help but wonder how Rabid Max might respond to his neighbor's raising wild grizzly cubs. Doug doesn't like the idea of transplants anyway, or the reintroduction process in general.

"Maybe it's not necessary," he says. "Maybe there are more bears out there than any of us realize. They've made it this far on their own. Maybe all we need to do, in terms of management, is keep our hands off. Besides, I have a real problem with taking a wild creature adapted to its territory and moving it two or three thousand miles."

Dennis nods. "Nobody asks the bear what it feels about being jerked out of its home up in British Columbia." When he says this, I'm reminded of the ritual the Blackfeet elder told him to perform whenever he trapped a bear on the reservation.

"Is that where a bear would probably come from, if there was a reintroduction?" Lavinia asks. Doug shrugs and shows his growing discomfort.

"We're not thinking in terms of bringing bears in," Dennis says. "Right now we're just out looking for sign and making field notes on the habitat. If I can get Round River started, I'd like to take a group of students into the mountains and do some quantitative work — mapping vegetation types, measuring habitat quality."

The cold beer is making Dennis garrulous. I've noticed that when Doug is about to say something brusque, something that might hurt someone's feelings, often Dennis will step in and keep the volume down, making his point firmly but in a manner more tactful than the words that are about to come out of Doug, a volcano of bluntness. It's more complex than good cop, bad cop. It's more like rivers and streams, or snow and rain. The force is the same, it's just a bit different in the telling.

"We're not going to bring animals in," Dennis says. "We've just volunteered to look for the bear, to bring Doug's expertise in" — Dennis doesn't mention his own — "and see what we can see."

Someday, whether we do or do not find evidence of the grizzly in Colorado, that will be the big question — whether to bring in bears from somewhere else.

Doug is on to something: his belief in the strong maternal culture of a mother bear and of a place — the way knowledge of a place is passed down, generation after generation, mother to daughter and mother to son, all the way into the frightening, challenging future. He calls it the culture of super-female, "Amazon" bears — these last, century's-end mother bears. It would be disruptive to take any female bear out of such a system in the attempt to shore up another. Then, too, there is the question of subspeciation, of the way the land shapes the animal through time — the effects the land has on characteristics within the population. The San Juans would have to start over with a new gene pool if "outside" bears were brought in.

We have another beer. Doug appears to be relieved that Dennis has finished with the subject of reintroduction. He asks Dennis Schutz, "Have you been seeing any mushrooms this year?" Doug's trying to be sly about it, but the distance from his heart to his mind is too short, and like a sieve. He waits for Schutz to say, "Well, just a few, not too many this year," and Doug smirks and says, "Uh-*huh*. And you're not seeing any of those lovely chanterelles, the beautiful apricot-colored ones that are so good when they're sautéed in a little butter and cream sauce, with maybe a pinch of garlic tossed in?"

And Schutz, and Lily and Lavinia, say no, they haven't seen any this year. Doug grins, leans back in his chair, and says, "Well, I'll bring some by the house before I leave. I found a secret place." His eyes sparkle, and he lowers his voice and looks left and right. "Never mind where. I'll bring some by when we head out of the woods." It means a thirty-mile detour, but no matter, his heart has spoken. "I'll just stuff a bag of 'em in your mailbox," Doug says.

The day's light is dropping. It's time to say goodbye. Lavinia wants to know one more time what our goal is. "Proof," says Dennis Sizemore, "of whatever is, or isn't, out there."

Doug asks Lavinia if we can take a couple of beers back to camp for George Fischer, and Lavinia says, "Sure, sure. Take them all." Our eyes brighten, and we begin stuffing our pants pockets with cans of beer. Lavinia laughs and Dennis Schutz, the old cowboy, nods approvingly, not so much at the beer drinking but at the simple manifestation of appetite, much as he might nod approvingly at a group of heifers or feeder cows moving in on a bale of hay.

As we drive out, there are more elk than before in the meadows. The cow elk are a beautiful golden color, and they move through the tall grass like queens, or ships at sea. There are a few smaller bulls coming from out of the trees, and we know that the really big bulls are still in the woods, waiting for dark before leaving the safety of cover.

We count dozens of elk, then hundreds, as we drive through Lavinia's valley. Too many elk, I say, and Doug and Dennis agree.

"The coyotes just can't do it," Doug says, shaking his head.

This reminds me of the fresh-killed elk that Lily found earlier in the summer, up around eleven thousand feet. She said her horse was frightened too, and almost unmanageable.

It wasn't a mountain-lion kill — those are too distinctive. This was something else, some scent her horse was not familiar with. Some scent that her horse's instinct, her horse's genetic data bank, warned her about: *Give this creature space.*

*

A grocery list, for when we pass through the little town of Chama. (We'll be passing right by Pagosa Springs, but Doug doesn't want to enter that swollen metropolis of five thousand people.)

Onions	Beer
Garlic	Ice (for drinks)
Toilet paper	Coffee
Rice	Stove fuel
Chilies	Goat

The store in Chama has everything but the goat, so we'll have to go meatless. Meat wouldn't have been such a good idea in bear country anyway. Or what we hope is bear country. It wouldn't have been proper to roast the succulent goat over the coals, juices dripping into spattering flames, wouldn't have done to lead the goat into the mountains — its bleating cries echoing off the canyons — only to kill it there on the bear's throne and eat it ourselves. We'll have to content ourselves merely to talk about the goat recipe Doug's had in mind.

It amazes us that there haven't been any livestock depredations. For a fact grizzlies lived for decades in these mountains without bothering sheep or cows. It's natural selection taken to the red line, perhaps; the grizzlies of the San Juans have learned not to come down from the mountains and not to prey on domestic livestock. But at what cost? It may be there is only one family of grizzlies left to exercise this hard-won natural selection. It's like whittling down a two-hundred-foot-tall Douglas fir to make a toothpick.

Doug wants to know if I, with my low-fat diet, will eat goat when we return to his place in Tucson after this trip. Will goat fit my diet? I tell him I will make it fit.

We arrive at our base camp in the lingering blue light of dusk. The days are growing shorter, as if being cut each day with a knife. Forty-Mile Ray and Jimmy Stearnberg and Big George Fischer have had a quiet day in camp, staying close so as not to make a lot of noise that might disturb the bears. They spent the time reading, napping, and eating, and taking short hikes along the river.

Goat or no goat, it's good to be back in the woods. We use George's wok to make a rice and chanterelle curry. Doug chops in pats of butter — "This is okay, right?" he asks me, and I lie and tell him it is — and more of the ubiquitous garlic, which seems to be growing in the back of his truck.

Now that we know where Dennis Schutz saw the four grizzlies nearly one year ago, in the morning we'll split into groups and make the long silent hike to that spot. We go through the story of Suicide Ned again — of how Doug and Dennis busted right in there and said what they had to say. We reenact the little pucker Ned's face made when Doug said, "We don't want your money, Ned." Doug and Dennis start giggling again when they remember how they burst out laughing the instant Ned went out to make his phone call. And even though it means we're broke, and are likely to stay that way for a while, we laugh until we're teary-eyed when Dennis recaps his final piss-away of the Round River funds: "Yes, Ned, in this instance I think Doug can speak for me."

We all laugh, not at Suicide Ned but at our own prideful foolishness. So the guy was a little late with the dough, so what?

It should be pointed out that some of the money did end up in a good place; it didn't vanish back into the pockets of the terminally rich. Ned gave the remainder of what was owed to one of Doug's favorite environmental organizations, the Wildlife Damage Review, in Tucson, which is a volunteer agency that keeps an eye on the government-subsidized killing of wild predators on federal lands. There's no telling how many coyotes and foxes that money has helped save. Maybe even a mountain lion or two. Or maybe only a songbird. No matter; the important thing is, Suicide Ned managed to put that money into the system it was destined to enter. Let the earth do with it what it will.

*

"The Wiseman bear," Doug says before going off to his tent. "Her parents and siblings — they fell through the cracks. No one ever saw or killed them, but they had to be out there in order for her to still be out there, twenty years later."

"The Platoro yearling," Dennis says — a litany of the dead. The little stuffed bear that Doug, Marty, and I saw last year at the lodge in Platoro had to have parents, and they were never found or killed, nor were any siblings.

We talk for a while about adaptability. About how Forty-Mile Ray's feet are in tatters, but still he pushes on each day, through the forest and across the rocks, a memory of the way he must have pushed on when he ran track over three decades ago. George is intrigued with our description of Rabid Max; he wants to know if it was a real machine gun, or just the kind you buy in a toy store.

"Hell yes, it was real," I tell him. "And his eyes, and the way he was rushing around — he was like a ferret. He was not at all happy that we believed in grizzly bears."

George crunches one newly emptied beer can and opens another, a one-two motion. He enters his FBI investigative mode again.

"Do you believe in grizzly bears?" he asks. "Do you believe they're up here?"

"Yes," I tell him, "I do." My feet and hands have gone numb again. We're at seven thousand feet, and the stars are silent and beautiful over the cold stony mountains.

"I do too," George says.

Morning. We move about quietly, briskly. The long dewy grasses are still flattened from the storm of a few nights ago. Steam rises from the trails as the sun labors to dry them.

The animals are getting used to our presence. Chipmunks have reappeared, scurrying along logs and chittering at us. Back in the aspen a hundred yards or so, we can see the shadows of elk moving cautiously up into the timber, returning from their nightly meadow feedings. The earth has tilted another click toward autumn.

We grunt and stretch and slip on our day packs, take gulps of cold water from the canteens, all without saying a word. Our day packs creak as we adjust them. Heavy boots, strong legs. Already Peacock's sweating. The day is different: not hot but bright, and the woods are silent.

Doug starts up the trail and we follow, passing through sun and shadow, and like no other day of my life, it seems, I can differentiate among the various qualities of shade cast by the trees — fir needles, spruce, and aspen. On my skin, the aspen shade is the most delicious of all. The day is electric, supercharged with sensitivities, with significance.

Departing the trail, we climb a ragged dolomite spine, tightroping it through a young stand of aspen. We see old bear-claw marks on a trunk, up high. We're ascending toward the sun and the faraway blue. There is no talk. We will angle over to the grizzly meadow with the breeze in our faces and the day's warming currents driving our scent past us and above, over the mountain.

It's all fairly theoretical, hence useless. A grizzly can smell an elk carcass at eight or nine miles, and after camping for several days, surely we are no less fragrant. What it will come down to mostly is luck and spirit, but still, we're going to do everything by the book.

Doug seems to fly up the steep slope. I find myself thinking again of the Sioux leader Crazy Horse — dead so young! — and of how his followers swore that he turned to stone whenever he dismounted his horse — transformed between the time he left his horse's back and touched the earth. There is a frequency to their testimony — gathered late in the last century and the first part of this one — that goes beyond the randomness of myth. I wish I could have seen it. Whether Crazy Horse really turned to stone or just appeared to — caught as he was in the air between the world of animals and the world of man — seems irrelevant to me. The fact is that Crazy Horse's spirit, his force, gave off that notion. Perhaps one day, people who hiked with Doug might say something similar: that every now and then he appeared to be flying up the mountains rather than climbing them.

Next year Doug will be fifty. Fifty years ago would have been a good time to see grizzlies up here.

My vision is rattle-shimmering again. I'm climbing fine, not even breathing hard, glorying just to be on the mountain, but when I pass into a patch of sun, spokes of light explode like a kaleidoscope.

Midmorning, we rest and sip water. We'll climb another five hundred feet and slide around south on the mountain, then split into two groups and approach the meadow from either side.

We've been spending a lot of time on the cliffs, picking our way up false chimneys and scrabbling up scree chutes. Forty-Mile Ray pulls his boots and socks off, examines his pink, flesh-torn feet in a patch of sunlight. Flaps of watery flesh are stripped loose where blisters have ruptured, revealing more painful flesh below. We grimace and look away. There's nothing to be done.

"I'll doctor 'em when we get back to camp tonight," Doug says. Ray stares at his feet a few seconds longer, pulls his socks back on, and then his boots.

A jet passes overhead, crushing the mountains' stillness in its wake. Like no other wilderness I've been in, the San Juans are plagued by overflights. Can't the jets detour left and right of this wilderness? These are real questions, not Crazy Horse questions, that need answering, and action. They need protest, need muscle, George's computer kind of muscle — the muscle of writing letters, not climbing mountains. Another jet passes over and Peacock curses, mutters something about "accountability."

We pause a bit longer, waiting for silence to return. A red-tailed hawk screams high in the sky. Even if the jets could detour ten miles south or north, it would help the grizzlies and the entire San Juan wilderness. At six hundred miles an hour, would such a detour cost anyone anything?

We start up again. Ray loses his footing and gasps as more flesh is peeled back. We lost our smooth rhythm, our flow with the mountain, when the jet passed, and for ten minutes or so we are clumsy and knock into dry branches or step on rocks underfoot. We're above the cliffs now, into the forest, moving through steep side-hill parks of immense Douglas firs.

After about half an hour we earn back our silence. The jet has been gone long enough so that it is as if a fever has broken. The woods settle back into their rhythm — the motion of small birds through the forest, and the trees themselves relaxing. The shade seems to grow cooler and the light on the ferns, filtered through the old canopy, grows softer again, and more significant.

Why would anyone come into church and kill all of the grizzlies?

Almost all of the grizzlies.

*

We're getting close to Schutz's meadow. The map shows that it's at the base of an avalanche chute and is about forty yards long and just as wide. We're less than a mile away. We stop and fill our canteens with water from the creek, then we split up. Dennis, Ray, and I will go up the shady wooded ridge on the creek side of the mountain, and Doug, George, and Jimmy Stearnberg will go up on the other side of the ridge. We'll be ascending toward some peaks, the conglomeratic pinnacles, which we can't see through the forest, but which look striking on the map, seeming to rise out of the forest like radio towers.

"From here on, no talking," Doug says. "Walk quieter than you've ever walked." He looks at us all fiercely.

Dennis, Ray, and I start into the dark woods, following a game trail. We have to move slowly to keep from popping dry timber with our boots. It's good day-bed country: a dark, safe place for bears to nap while waiting for the safety of night. They lie curled up in their beds and sleep through the heat of the day. Sun-warmed breezes carry all scents up the mountain to where they rest in flattened areas with the sticks and branches scraped away, and sometimes with grass carried in for padding. We stop and look at some ancient moldering bones that stick up from the moss and dark mulch on the trail's edge. Years of twigs and fir needles and forest rot have covered the bones. Without a word, we pull up a few of the bones to determine what kind of animal it was.

An elk. We piece the scapula and a femur and tibia together and place a few ribs back in their proper position. We spy the skull, with antlers still attached — he probably died twenty or so years ago — and reverently, we pull that from the soil too. It's a huge royal eight-point, the only one I've ever seen. Without saying a word, we pass around the great skull and admire it. Moss has grown on the bald pate, and the antlers are worn, whittled by the teeth of squirrels and porcupines. The old boy died in the fall, and on the trail. Almost certainly he fell to a predator — more noble, perhaps, than succumbing to winter weakness and starvation. We lay the head where it belongs, near the scapula and some stray vertebrae — as gently as if laying a favorite old dog to rest — and continue on our way.

Off to our right, through the dense tall trees, we can see the bright light, one of the many avalanche chutes plunging down from the mountaintop, each chute giving direction and force to unnamed waters.

All through this shady little basin there is the smell of bull elk in rut. The bulls have by now herded the cows down from the high country. I find myself imagining that conception has already occurred for some of the cows, and that — even as we walk through their woods — the cows have more young bulls and cows spinning in their wombs like astronauts: one cell, two cells, four cells, eight, and on and on.

Dennis is walking ahead of me, Ray behind. We still haven't said a word, but I know from the way Dennis is stopping every few feet and scanning the woods that he is looking for day beds. The shadowy forest rises steeply before us. Up ahead we can see the ridge, the narrow spine, on the back side of which is the grizzly family's meadow. The slope rising to the ridge is parklike, with Douglas firs so tall and thick that they prevent sunlight from dropping all the way down. Between us and the ridge there are square dolomite boulders the size of small houses, and lichen-colored — the whole area like a meeting place, ceremonial and strange, for druids. The boulders must have tumbled here long before these firs grew old around them.

We move up the trail more cautiously than ever, a step at a time, and are aware of something, some kind of compression, as if we're pushing against something, forcing it up. Something different from the real world — or the one we have become accustomed to.

Bears. Bears ahead of us, clicking their teeth and grunting, a sound unmistakable to me after having been in grizzly country up in Montana, with Doug. Dennis hears it too, and stops. It's coming from behind a boulder straight ahead of us, up about fifty or sixty yards, right next to the trail.

We freeze and listen to an angry bear. A deep, almost subsonic sound. Surely these must be the bears seen by Dennis Schutz. We stand like pigeons in the middle of the trail. I have my camera out and ready, but Dennis — the linebacker — is blocking my view. It occurs to me that the bear might charge us, or rather, might try to flee downhill, and run right through us in the process. We've gotten too close.

Even if she barrels into our midst, I'm determined to get pictures of her. I click the autofocus to *A* and bring the shutter speed down to one sixtieth of a second, to account for the dark woods. I stop breathing and level my camera on that square boulder.

We listen to the grunting for ten, maybe fifteen seconds. A breeze stirs the trees to our left and carries our scent with it, straight up the hill.

There are two or three quick exhalations — *whoofs* — and then the crashing of sticks and limbs and some squeals, pain and anger mixed, and then a few more sticks breaking, and then silence.

The ridge is eighty to a hundred yards away, and though it curves around to our right, I feel sure that in an instant we will be privy to one of the most magical sights in nature: grizzlies. But what if the bears run down the other side of the ridge and bump into Doug, George, and Jimmy Stearnberg? Again, I hold my breath.

Five seconds, four, three, two, one . . .

Another cub squeal, and one more brief timber-smashing sound off to our right. Then silence, and absence, and a light breeze high in the tops of the trees. The wooded basin feels inexplicably hollow.

As if rooted, we wait, still not knowing whether we will or will not be charged.

I distrust my senses. I can't believe the bears went without all the usual crashing of timber that you hear when you jump up a deer, an elk, or a moose. These bears — with the exception of the two brief crashes, one straight ahead of us and one off to the right — seemed to *flow* away.

We wait and wait, but the absence, the emptiness, only grows stronger, and we know that they're gone. They've crossed the little ridge without our having seen them — having used the big boulders to shield themselves from view.

They were unmistakably bears, though — the tooth clicking and the grunting and the squeals of young bears being cuffed. And this was within seventy yards of Schutz's meadow.

Were they grizzlies?

I think that if they had been black bears, the mother would have sent them up a tree. But grizzly cubs, three- or four-year-olds, might be too big for tree climbing. Already their dagger claws might make it easier for them to run.

Dennis moves carefully up the trail. Despite the ringing emptiness, the sense of abandonment, we're not entirely certain there's not a bear or two still holding out, hiding behind the largest boulder — a three-hundred-pound cub, perhaps, quivering in fear.

I follow with my camera cocked, ready to get a picture of anything, even a fleeing brown rump. When we are within ten yards of the boulder, we stop and whisper, "Bear? Hey, bear. Hey, *bear?*"

There's no response of any kind. We fan out about the rock and search for hair and scat (a napping grizzly will often defecate when startled into flight). We search for a day bed, but find nothing.

In retrospect, it seems that we did not look hard enough. We might have been off by a mere ten yards but, flushed and excited, we attended only to the things we expected to find — a big scat, or a big day bed. After a minute or so of searching for these things, and not seeing them, we took a game trail that went up and over the ridge to our right — the second crash of brush came from this direction — and we followed this trail, looking for fresh scat.

These bears did not get to be survivors by panicking and leaving cover when intruded upon. They simply went in deeper.

We hike to the ridge and follow it back to the promontory rocks, to better scan the area — those eerie spires that straddle the little ridge and rise above the forest like strange sculptures. The Hoodoos, where we are supposed to meet Doug and the others.

The ridge has a wide and well-worn game trail running along its spine. It is evident that bears using one of these trails would make no sound at all. Below, on the back side, we can see the small sunlit patch that is the grizzly meadow.

Doug, George, and Jimmy are somewhere below that, but we're pretty sure the bears would have scented them and run away. For a moment, though, I consider this ironic possibility: Doug has been injured or even done in by a grizzly bear that is not believed to exist. Most ironic of all, Doug has done nothing to deserve his fate. We, his friends, are responsible, having startled the bears and pushed them down in Doug's direction.

Surely we would have heard their yells if there'd been trouble.

Dennis, Ray, and I sit down in the deep shade at the base of the largest Hoodoo and share a small box of raisins. I enter these notes in my pocket notebook:

I thought she would've sent them up a tree.
The thrill, 11:15 A.M.
I believe they're here.
Upwind. The squeals of fear.

There are a few tall aspen trees mixed in with all the dark firs, and their leaves chatter softly in the breeze. We grow strangely sleepy, sitting here on the magic ridge, the magic spine. Perhaps it is an overcorrection from the surge of adrenaline we took on not forty-five minutes ago, or maybe it's the spell of this one spot, because soon we are stretched out and falling asleep, even snoring lightly in these peaceful woods. It seems odd to go to sleep after having flushed grizzlies. I can't explain it.

If the bears were to calm down and return to their day beds in the basin, they would probably come right down this trail and step over our sleeping bodies.

We nap as if touched by angels, and when we awake, we do so slowly, sitting up as if we've been asleep all winter long. There's still no sign of Doug and the others, but this is where we agreed to meet, and so we wait here.

But after a while I can't wait. Those bears ran uphill. I want to find sign. I want more. The mountain rises steeply above us — we're at about nine thousand feet, but the peak rises to above twelve thousand feet. I watch a pair of ravens drift across the sky. They're cawing, making adjustments and corrections in their glide. They seem to be following something all the way up the avalanche chute, the chute we'd been ascending when we jumped the bears from their day bed. The ravens are giving short, inquisitive caws, and I feel certain they're tracking those four bears, seeing what all the fuss is about. Nothing happens in the woods that the ravens don't know about.

If that peak above us was the grizzlies' haunt in the sixties, then perhaps it is still their haunt now. Two plus two equals four, I think greedily, wanting still more sign, more bear.

What if we *had* bumped the bears into Doug and his party? I remember Doug's "Feed the Bears" card he keeps in his wallet, and know that it contains his true wishes, if it came to that. If history repeats itself, I wonder if this was how it happened for Niederee and Wiseman. Did Niederee force the bear (perhaps with the aid of an arrow through the lungs) from cover, out and over the top of his partner? What would it be like to be hit by five hundred pounds of fur and teeth?

I know that either way, dead or alive, Doug would want me to keep looking for sign. And if the bears had *not* run over the top of Doug, who was somewhere below, they would have gone higher, seeking sanctuary.

I write a note in my notebook and attach it to my backpack, which I leave leaning against the Hoodoo rock. Dennis and Ray have closed their eyes again.

12:35 Doug — We heard bears at 11:15. If I don't meet back up with y'all, don't worry, I know my way back. If I can find a covered approach to Z Mountain where I won't blow anything out, I may go up there. — Rick.

Just as I sign the note, Doug comes walking up. He'd moved so softly through the trees that I did not see him until he was almost before me. He is alone, as quiet as a ghost, and that is my first thought, that the bears got him and this is Doug's ghostly spirit roaming the Hoodoos.

I start to say something, but he puts a finger to his lips, a silent *Sshh*. Too many words, once again.

He walks down the trail to where we dropped our packs. His eyes are larger and rounder than usual, alive with a secret, a big secret. I wonder if his eyes have been scorched by the grizzlies racing in the woods, running right past him, coming so close that he could have reached out and touched them. Then I wonder where Big George and Jimmy Stearnberg are.

"What did you see?" I whisper to Doug when he is closer. "Did you see the bears?"

"I found a freshly overturned rock," he says. "It was a big rock. I think they've been in here digging."

"We *heard* them," I whisper. "We jumped them from their day bed. They ran off that way." I point up the mountain behind us. "I was afraid they were going to run right over you."

"Did you see them?" Doug asks.

"No. I don't know how we couldn't have, but we didn't. There was a rock in the way. But we heard them. They were making that tooth-clicking sound like we heard when we went into the Grizzly Hilton, and then we heard these squeals, which I guess was the sow slapping the cubs, telling them to run."

"Did Dennis hear this? Was Dennis there?"

"Oh, yeah. Dennis heard it too."

"Damn. They're here. Listen, we have to be real careful. We have to be real quiet."

I can barely hear him. The breeze in the treetops muffles Doug's whispers.

"This is a real strong place," he says. "That boulder was just turned over, fresh. Maybe last night, maybe this morning." Shafts of light work their way down through the trees in places, the way light falls from high skylights in an old train station.

Doug steps past the slumping forms of Dennis and Ray, goes up to the largest Hoodoo, leans in and spreads his arms around it, places his ear against it as if listening to something inside. He smiles, then takes his baseball cap off, as if remembering his manners. He stays in that position for a long time, feeling the mix of pebbles and gravel and river rocks press against him.

Every place on the earth has a consciousness, an awareness of being that is the sum of its existence. You can feel this in New York City as much as you can feel it in a Utah desert or a Montana wilderness. But I've rarely felt it as strongly — that sentience of place, a consciousness beyond our own — as I do here on this wooded ridge.

The bears got to it first. They found it, and have been living here all along — through the last several centuries, anyway. There are probably other such places, small good places, in the San Juans, and they need protecting. More important than any legislation, though, they need and deserve our reverence. As a culture, we're big on the quick fix, doling out money but stopping when it comes to doling out our hearts, our understanding. It's a fairly easy act to "create" or protect or "set aside" a single wilderness, but if we do not give it our respect, we have not fixed or even approached fixing the problem.

Without a doubt, the San Juans need more space, more of a buffer

against the iron edges of civilization. But just to throw up more bounda-
ries is not a real answer. It is a partial answer, a stalling tactic, a delay
against the loss of wildness. For wildness to survive, for wildness to
return, reverence must also return. Not so much knowledge, but more
understanding. Respect, awareness, caution; providence, prudence, com-
passion — what sounds like a shopping list for the Quakers, perhaps, is
really more of a checklist of the wild.

The autumn sun is warm, even in the sliver shafts with which it's
being delivered to us. Dennis and Ray stir, and Doug turns away from
the rock, walks over, and prods them with his boot, and they slowly
open their eyes.

Dennis tells Doug the story of the bears. What I remember most,
besides the sun and shadows, is the seriousness of the moment, the
responsibility. It was a thrill and a joy, but it also had a weight to it. It was
a weight — almost a burden — that had not been there when we left
camp that morning. We would no longer be able to say, "Ah, we're just
up here in the woods fucking around and having a good time."

The bears had entrusted us with a secret.

George and Jimmy Stearnberg had gone back down to camp — had
split off from Doug early, with rumbling stomachs and low-grade
fevers. They suspected giardiasis, but it could also have been Dickelosis.
Doug picks up my notebook, studies my note, and nods, says that it
would be a good thing for me to climb Z Mountain. We'll send Dennis
and Ray across the meadow and up into the dark steep woods on the
far side of the next avalanche chute. Doug will go back down and
explore the lower woods around that freshly overturned boulder.

"Be back by three," Doug says. "We'll head out of here then. I don't
want to be around here anytime near dark, which is probably when the
bears'll be back. We want to be sure and keep from blowing them out
for good."

This means I have about an hour to climb three thousand feet and
an hour to come down. I'll have to hike crisply.

I jettison everything — water, backpack, food — and start up the
steep game trail with nothing but my T-shirt and pants and hiking

boots, feeling set free, feeling like a greyhound sprung from a cage. I head up almost at a trot. The electric sparks in my vision are forgotten — so what if my toes are asleep? Maybe it will stop there; maybe the rest of me will never go to sleep.

The ravens are calling again from that same avalanche chute, so I angle in that direction. I am hiking the wooded ridge for no more than five minutes when they sense me moving through the trees, and they wheel around and fly down to where I am. They alert the world to my presence, with short caws that sound as if they're giving out information about me: name, height, weight, date of birth, Social Security number.

After I've been processed, the ravens drift back to the headwaters of where we surprised the bear. I keep climbing, heart pounding, up to ten thousand feet now. I am up above the Hoodoos. I look down on the treetops below, see those rock pillars jutting above the trees, rising out of the forest like stone pistons, pistons for the engine of the world. I know that Doug and Dennis and maybe even Ray are looking up at those ravens and listening to them, and they know exactly where I am. I imagine I am a bear. It seems clear that if everyone below knows where you are, and you do not want that to be known, then all that is left to do is to climb.

The higher I go, the drier it becomes. Crossing an open patch of dry grass, I jump large numbers of grasshoppers that are in a hurry to eat the grass that's almost hay now, in a hurry to eat it before the snow can bury it. What bravery it must take to be a grasshopper at this time of year, when afternoon can bring dusty heat and evening, frost and snow.

Across the valley I see the jagged spine of the Continental Divide, level with where I am, closing in on eleven thousand feet. These bears were on the Pacific side of the country, but within a day they could cross that valley and be on the Atlantic side.

The sun and wind and the thinner air dry me out, and I half trot, sometimes up a pitch so steep I have to hold on to tree trunks and branches. It's too bad that snow's not over here, as it is on the Divide.

Maybe in two or three weeks it will fall here, but by then we'll be back in our other lives. That late in the year, and up this high, the bears will be entering dreamtime, crawling into their dens.

Do they all den together? Probably not, but perhaps. These bears are almost certainly atypical. The land will ultimately shape the animal, will mold the species, though it is a partnership: the animals help shape the land. And perhaps these last holdout bears, the uncompromising ones, have taken on the nature of the elk, are more of a small band now, under the guidance of the dominant matriarch. If this is true, it is all the more tragic that the last known bear killed, the Wiseman bear, was a mature female.

I recall reading about the bait stations the Game and Fish Department used in the 1960s, the horse carcasses. I think of a sad old nag being led up the tortuous scree of a mountain, being given a gulp of water — water the quivering horse had hauled up himself. Then the crack of a pistol, and the angels, the buzzards, descending. Then the grizzly consuming the horse, taking up horse flesh and horse spirit and turning it to grizzly flesh. That grizzly was never seen by man, but it happened, once upon a time, in these mountains.

*

I take long strides and find myself lifting over a false ridge that I thought would be the crest. Instead there's only a grassy basin and more steep mountain ahead of me. Even this high there is still an occasional game trail, with deer and elk droppings, tufts of fur stuck to the sap of a fir tree. It's wonderful to think of such large, powerful animals moving around at the top of the world.

Occasionally I come across a marshy seep, a bright swath of green sedges, wildflowers, grass, even algae. I pause briefly to check for tracks, finding none. What animal, save a man, would dare damage such delicate springs by trampling and wallowing in and fouling them, shutting off the source? Even the elk have stayed out of this spring, though I am sure all the animals drink from its trickle, which is so thin that it is only an iridescent shimmer.

There's dwarf vaccinium everywhere. Ahead, a wall of sheer cliffs rings the mountain's top like a crown. More springs glisten against the face of the mountain, where the cliffs meet the pitch of slope. *Where is the sign?* I write in my notebook. I'm altitude dizzy, and pause to suck in the thin air, to relish my heart's torment. *Show it to me.* I feel that I'm right upon it.

I look a little farther up the slope. I'm sweating, but the high wind dries me immediately. My eyes stop on a large gleaming white bone. Is it an ancient leg from one of the bait horses? I walk over to it. It's a rib bone, and then I find a femur and vertebrae. I search for a skull, and find it. Elk. *How did you die?* I look through the grass for more bones, feeling the wind wash the sides of my face. Balsam root is growing at the edge of this cool steep meadow below the top cliffs. The earth is a miracle!

I'm heart-fluttering: at this altitude, your breath doesn't come right back when you suck it in. The mountaintop, or perhaps the sky, holds it from you for a moment longer, as if to get your attention — as if to make you look around and see.

Juncos feed in the firs all around me, close enough to touch; they're unafraid, as if I'm a junco too, as if the whole world is made of nothing but juncos. Their actions, the cracking of seeds and even the flutter of their wings, echoes *autumn* across the high country. There are chipmunks running through the tall grass, cutting stems so that the seed-heavy tops fall over like trees crashing in the forest, and the chipmunks fill their cheeks with seeds, gorging like little bears. Shooting stars are beginning to blossom in and around the lush meadow. Spring has only just this week reached this elevation, and next week will be yielding to a moment of summer, and then a brief fall, and then the long winter.

To the crest, then, which is long and skinny. All of the physical world's below me now, every last inch and ounce. If there is anything left in the world it is invisible, just above and beyond me, in the sky.

The trees have trickled out into windblown, runty things. I walk over to the other side and look at the slope below — deep dark timber, warmer temperatures, bigger trees — then look off at the rest of the blue and white San Juans, ice-capped and jagged, spruce-colored. Infinite space. If there are four bears on this mountain, then surely there can be more even farther in, between here and that jagged far horizon.

It's grassy up here. There's a small cairn, no telling how old, the kind of thing people like to build with rocks when they come to a place like this: some vestigial human gene governing territoriality. I look down over the cliff wall, thinking it would be an ideal place to make a den, either in the earth or in a limestone cave. I look for freshly dug earth. Perhaps the four blond bears are still down in the timber, hiding between here and the Hoodoos. I watch the open side-hill parks below, hoping to see them amble out through the grass, nipping at it like elk.

Above twelve thousand feet, I find the bear scat. It is off to the side of the elk path that runs along the skyline. The scat is the largest I've ever seen, only about seven or eight inches long but thicker than my wrist. It's dry and weathered, with grass growing around it. When I pick it up, I see that it's not old: the grasses beneath the scat haven't yet grown into and through it. There's bare soil beneath it, from where it inhibited the grass's growth with its acidity, its nitrogen. Definitely this year's scat, probably from May or June: it wouldn't survive a season of weathering, not at this altitude.

I put the scat in my baggy side pocket; it is as light as a hunk of French bread. I walk the ridge farther north, looking for more scat and looking for those bears, now that I know they are below me for sure, on one side of the ridge or the other. To the east, down in the basin out of which I have just climbed, something catches my eye, something coming out of the woods and across the edge of the grass. Something blond, something galloping. I crouch behind a boulder and watch, believing.

One blond creature, a huge one, then two, three, four, five. A herd of cow elk. It's the height of the rut and they're acting all skittish, looking back over their shoulders and trotting. Why aren't there any bulls with this large herd? The females keep coming out of the trees, eleven of them now. They cross a tiny spring, run up a steep slope in a panic. I imagine that bulls are back in the woods fighting one another, though

I can hear nothing. I'm about three hundred yards above them, with the sun at my back, the sun in their eyes. They are a light yellow, with a beautiful glow. I crouch there with the immense grizzly turd in my pocket and can almost *feel* the bears. I feel that if I just wait there another hour, or until dark, then one of the grizzlies will come loping out into that open meadow, followed perhaps by another, and then another, and even a fourth.

I look at my watch and realize I have overstayed: it is time to meet up with my friends and get out of the bears' woods before twilight. I rise and move along the ridge, staying low, and trot across the grassy plateau, which is almost the size of a small airstrip. I find my chute and plunge down. In an hour I'm back on the Hoodoo ridge, walking wobbly-legged across that dark pass between the two avalanche chutes, with the little meadow below me on one side and the dark day-bed timber on the other side. I'm actually ten minutes early, but Doug and Dennis and Ray are packed and waiting when I come down the trail.

Doug is anxious to be out of the bears' home. I'm anxious for a sip of water, but first we need to go through the ritual of sharing the day's stories of the hunt. It's like returning to a wolf pack. They want to know what it was like up there and what was on the other side. I want to know if they heard or saw or found anything else down below.

Doug's eyes have mischief in them when I ask if they found any-thing. I notice for the first time that he's peeled off his outer nylon pants and is wearing just his light khakis. He points down the slope to what looks like a scarecrow. It's a bulging headless man wearing Doug's clothes; lumpy, as if stuffed with newspapers.

"Chanterelle Man," Doug says. "The mother lode. Chanterelles all through this forest. More than I have ever seen. More than I have ever dreamed."

"Chanterelle Man probably has a street value of a thousand bucks in New York," Dennis says.

But we're not in New York. The afternoon is finally starting to turn from hot to cool, that shortening cycle of the seasons, every afternoon a bit cooler than the one before. A Steller's jay flies through the trees, passes between us and the sun. It twists its head at us as it flies past, but it does not betray us.

Four men, four bears. We have to believe that more good than harm will come from our being here: the dissemination of this news, this testimony from the heart of the country. If by our coming here the secret will now be believed and respected, then I hope that it was worth it. But without a doubt it is now time to leave and maybe never to come back.

"Pass me some of that water," I croak. Dennis hands me the canteen, which is half empty, and is all we have among us. We must make it last until we get back down to the creek. Once we start, we will move quietly and won't stop until we've given the bears their space back.

Doug's nearly euphoric, as if some ripcord has been pulled within him, a ripcord of the wild, allowing him to inflate with a silvery, lupine wildness. We're sitting there in an autumn light, preparing to leave this spot to which we may never return, when we see animals moving toward us below, on the meadow side of the pass. We can tell right away that they're elk — bull elk, darker than the yellow cows. They're tiptoeing, sneaking away from something. We sit motionless and watch as they pass through the woods between our ridge and the meadow.

They're not thirty yards away, but so intent are they on sneaking away from whatever's behind them that they don't have a clue that we're that close to them. The sun shafts cast watery light on their bodies as they move through the forest, four big bull elk, and perhaps they are moving away from the bears, or perhaps they are moving away from a larger bull. We sit as still as statues and watch the sun glint off their antlers as they flow past, slightly below us, now only twenty yards away. Their black hoofs click on rocks here and there, but their secret, hurried passage is a quiet one. When they have gone past us, it feels as if they have resealed the sanctity of the woods. Four bears, four elk, four men.

I show the others the big bear turd I found. Doug reaches for it as if it's made of gold. "That's it," he says in quiet wonder. "That's it." He hands it to Dennis, who also examines it.

"It's a grass scat," Dennis says, "an early summer scat. Look, it's got some hair in it."

I hadn't known the hair would be that fine and wispy — more like a spider-web strand than the coarse deer and elk hairs I'm used to seeing. The hairs are silver, almost white, like an old man's long hair.

"We'll send it to the lab," Dennis says. "We'll see."

A cloud passes over the sun, and it suddenly seems as if we've overstayed our welcome. We rise and shoulder our packs. Dennis puts the big dry scat in a sample bag, and Doug and I each carry one half of Chanterelle Man. Doug's knotted the sleeves and the ends of the pant legs to keep the mushrooms from spilling out, and I place Chanterelle Man's lower half around my neck and over my shoulders, wearing him like a heavily padded life preserver. Doug carries the upper half like a dance partner, and we're off.

*

In camp that night we empty our booty onto the hood of Doug's truck, mounds and mounds of sweet moist chanterelles, a pungent smell that's part fruit, part meat, part sex, part forest, part wine, part flower. Doug's happy, but he seems a bit off balance. Months later he will write about how the joy of the day's bounty was mixed with sadness: how Chanterelle Man reminded him of the body bags used in Vietnam.

By firelight we string the chanterelles onto huge necklaces made of fishing line found in the depths of Doug's truck, and drape the necklaces over tree branches to dry. We sauté and eat as many mushrooms as we can, washing them down with George Dickel. I'm remembering the simple elegance of Aldo Leopold's Round River diagram — as published in his book with that title — the notion that inspired Dennis's idea for his school. With a stick I sketch it in the sand:

rock → soil → oak → acorn → squirrel → Indian → soil →rock →

In this pipeline of life, the highest achievement anything can reach is the soil itself. Which becomes, of course, the beginning once again.

Leopold writes of how some of the river's elements — oak, acorn, squirrel — return to the soil prematurely, eddying out of the system a step or two early: the acorn not gotten by the squirrel, the squirrel not gotten by the Indian. "Owing to this spillage enroute, only part of the energy in any local biota reaches its terminus." And, he adds, "In addition to losses from spillage, energy is sidetracked into branches . . . Nor is food the only important thing transmitted from one species to another. The oak grows not only acorns; it grows fuel for the Indian, browse for deer, hollow dens for raccoon, salad for June beetles, shade for ferns and bloodroots. It fashions domiciles for gall wasps; it cradles the tanager's nest; . . . and all the while its roots are splitting rocks to make more soil to make more oaks."

It occurs to me here, doodling in the dirt — *mountain, snow, sun, chanterelle, bear, man* — that this is where we are in the river, this night, around the campfire. We are in the "spillage" part, Doug and Dennis and George and Jimmy and Forty-Mile Ray and me, and it is precisely the spillage that gives meaning and richness to the way a life is lived, to the flow of the river.

This ability, this potential, to "eddy out" of the pipeline for a short period is one of the real and glittering wonders that a large wilderness has to offer us: something to distract us from, or instruct us during, our kicking return to the soil. A way to remember our history.

We pass the pot of chanterelles, we pass the Dickel. The bears clicked their teeth at us today and ran away. It could be a century or two ago, six men feasting on buffalo ribs. The delicate mushroom necklaces hang from branches all around our camp, jewels for the soul, a nice surprise for any deer who might wander through the woods tonight. They're welcome to share.

*

One more day spent in the lowlands, far below the bears. Doug wants out of the area entirely, to let the snow come and bury everything and give it all a rest, the same rest these bears have had every year, perhaps since the Ice Age — twelve thousand years and now one more. But we're waiting to meet Jim Tolisano, so we need to occupy ourselves in the mountains for one more day.

We hike quietly through the bright sun, moving along the creek combing the woods and thickets for signs of bear, jumping elk with every step, it seems. We work in a circle, in two groups, at the base of the mountain. We find a few old claw-marked aspens, and who can say, grizzly or black bear? The claw marks are pretty, their curves reflecting the musculature of the bear's shoulder attachment — a curved pulling sweep, like the arcs of a circle or the bends in a river. A Steller's jay cries out from the woods. Doug tells us these birds are great at imitating the cries of golden eagles.

We back farther away from the mountain, giving the jay more space, giving the bears more respect than ever. That's one of Dennis Sizemore's Round River concepts, and one of the tenets of conservation biology: to have cores of wildness around the country, with concentric rings radiating from those cores where man's activity is gradually allowed to increase outward. It's a more integrated approach, more organic than the outdated practice of industry and community slamming on the brakes at the wilderness area's boundary. The concept is gentler on both the towns and the wilderness areas.

It's not a complex model that Dennis is proposing — it's already being used for single-species management, such as Yellowstone's last few grizzlies — but it's a hell of a lot more sophisticated and logical than the current model, the brick-wall approach. The new way works like a stone tossed into a pond, with bands of wildness.

"We see, then," writes Leopold, "that chains of plants and animals are not merely 'food chains,' but chains of dependency for a maze of services and competitions, of piracies and cooperations. This maze is complex . . . Paleontology discloses aboriginal chains at first short and simple, growing longer and more complicated with each revolving century of evolution. Round River, then, in geological time, grows ever wider, deeper, and longer."

We must be able to accommodate deep cores of wilderness within our country, from which to draw inspiration and energy for our own strength, our own wildness. The wildness against death. The wildness *toward* death. The wildness that is life.

*

Tolisano arrives in camp shortly before dusk, walking up the old logging road through the aspens. He couldn't make heads or tails of our directions, but had been able to follow a thin wisp of campfire smoke against the darkening sky. He abandoned his car, an ancient Toyota, in a mud puddle about a mile away.

He's a small man who reminds me of a Greek Kit Carson (Tolisano lives outside Santa Fe, Carson's old home). Black-haired and dark-eyed, with a short-trimmed black beard and a broad canvas hat held on with a leather chin strap, he seems to be more whole in the woods, more at home, though how I can tell this, having never seen him in a city, I can't say.

I sometimes expect small men to be tightly wound, coiled with energy — especially Tolisano, who's in the midst of environmental and cultural work for the governments of Papua New Guinea, Malaysia, Borneo, and Costa Rica, and who has other projects, other passions, in his sights. And if Dennis can get the Round River program started — if he can come up with six students to bring into the San Juans next summer and teach them how to look for bear sign and how to map diversity — then Tolisano will handle that project, too.

I don't know the man, but already I worry about his family life. How many things can you love? Can you love *everything* that is wild? I believe that you can, and I like to believe that it all strengthens rather than weakens you.

I look at Tolisano's quiet peace — he's standing around shaking hands with us all and staring up at the beauty of the yellow aspens and cottonwoods as the dusk slides in — and I realize that in this case it doesn't really matter whether this project, the story of grizzlies holding on in Colorado, drains you or strengthens you. In this instance, it's not even a question. The pilot tells the paratroopers to jump, and you jump. There's no time to weigh options, no time for pause. There are few left in this century, and one of our comrades in the wild is down.

Never leave a man, or a bear, behind.

The last night, the last campfire, and there's an edge of loneliness in camp, with autumn pushing us out of the mountains and back into the world. The chanterelles are still delicious, but we're low on everything else, including whiskey. Forty-Mile Ray will be heading out tomorrow, back to Michigan with tattered feet, and I may never see him again. It may be another twenty-five years before Doug sees him again; add twenty-five more to that, and there won't be a next time. Jimmy Stearnberg is heading back also, back to city life in Denver.

Our hearts have already told us what Dennis and the pathologist will later confirm, or punctuate, when they look at the scat sample's hair under the microscope: there are grizzlies still living in Colorado. We tell Tolisano the news and he's blown away. It's more than we could have hoped for, and so early into the search, too. But we don't know what to do — don't know what the bear wants us to do. Doug's instinct, and Dennis's, is to hold the secret. To let winter come, let the knowledge be knowledge, let the mystery be mystery, without asking for any land management changes, any brick-wall stuff.

We can start the dawn of consciousness — that there are still grizzlies in Colorado, can even prove it now — but we cannot change ethics or ideas, not with a phone call and not with a press release. So we'll just sit tight for a while. It's a little deflating to realize that after pursuing the proof so passionately, and then finding it, we can't really do anything with it.

The mystery of grizzlies in the San Juans is still at least as important as the knowledge, and so we sit around our campfire and turn our secret over and over, not knowing what to do with it or why the bears revealed themselves to us or how we can help them. We turn the secret over and over in our minds like a raccoon washing an oyster pearl in a stream, examining it.

Although we can only sense abstractly the coming problems, we understand that those who don't want to believe simply won't. I fear someone will accuse me of planting the grizzly scat, and of course, later on, this implication is made. Later, too, researchers will take DNA samples from our scat's hair and from the hide of the Wiseman bear to show a genetic connection, which will answer that accusation. But by then the argument will have shifted elsewhere — those who do not want to believe cannot be made to believe. It's true we need stronger proof — a video clip best of all. (Even the proverbial dead bear wouldn't work, because then agencies could say, "Well, that was the last one; *now* they're all gone.")

The last thing we want is a starry-eyed rush of backpackers in the San Juans, all trying to "help" the bear by finding grizzly shit. And yet we want the people in charge — not just Congress, but the local communities — to understand what we have, what we are still blessed with.

"When we see land as a community to which we belong," Leopold writes, "we may begin to use it with love and respect."

It's going to be stepped up a notch from here on out. It's not going to go away, this dawning knowledge. We have found bears. We have confirmed what Dennis Schutz found. And someone, sometime, will probably set up a heat-sensing video camera and confirm what we have confirmed. And then someone else will repeat the process, may even try to capture and collar one or all of the grizzly family.

We sit around our fire and turn the secret over and over, examining every angle. We pass the bottle, we tell stories. It takes strength to hold a secret. And to release the knowledge at the right time. (Doug will do this at a conference in Boulder, later the next year.)

We begin, one by one, to drift off to our tents, ducking under the garlands of chanterelles. Finally Jim Tolisano, George, and I are left, staring at the flames. Jim has just come into the woods and is not ready to quit the night yet. George is at the other end of the spectrum, hanging on to every last second in the woods before he returns to Salt Lake City tomorrow.

"A beer would sure be good right about now," George says. He looks up at the stars, then across at Jim. "You want to ride in and get a beer with me?"

Jim smiles, shakes his head, reaches in his pocket, and hands George — this big man whom he's never met before — his car keys.

"It's back about a mile that way," Jim says, pointing through the forest.

George sighs, looks up at the stars, thinks about it, hunches his shoulders, and stands up. He pulls a flashlight from the depths of his parka, flicks it on to test it, looks up at the stars again, and walks off into the night.

Jim and I listen to the brush cracking, and then the jungle swallows him.

✷

"He must really have wanted a beer," Jim says a couple of hours later.

A shooting star whizzes right over our heads.

"I think he was feeling a little out of sorts," I say. "You know how sometimes it does you good to just get in the car and go, to drive with the windows down?"

Jim nods, but I can tell it's been a long time since he's felt that way. Like Doug, he was on the road most of last year. I can tell that what does him good is to stay in one place. To poke a stick into the campfire. To stare at the flames.

Another hour has passed, and now it's midnight. "I hope he hasn't run out of gas," Jim says. We're just staring at the fire, or what's left of it. "It's only about thirty miles to Chama. He couldn't have gotten lost."

At twelve-thirty we hear the far-off sound of an internal-combustion engine, an intrusion in the night's stillness. The sound has an inescapable death-stutter to it, and it echoes off the valley walls.

"What the shit is that?" Jim asks.

"Probably a logging truck," I say. "Sonofabitch is stealing government logs at night — highgrading bastard."

The engine sound fades and the night's peace returns. Then, at one A.M., we hear branches cracking, and a mud-splattered George steps out of the woods. He's carrying a paper bag, and he walks right past us to the stump he'd been sitting on before he left. He sits down heavily, pulls a six-pack out, tosses the paper bag on the coals. It bursts into flame. The firelight gives us a quick look at his suffering face. He pops open a beer, takes a long swallow, and offers us each a can, looking directly at us for the first time.

"I wrecked your car," he tells Jim.

I like the way no one in this crowd beats around the bush.

"I tore it up bad." George finishes the beer with his second swallow, opens another. Neither Jim nor I have accepted his offer of a beer — we lost our craving several hours ago — but for George the night is young, and misery has him locked in both her arms.

Jim watches George watch the fire. Is he thinking of how far he is from Santa Fe and his family? Always, it must seem, he is away from them. Is he now contemplating walking home?

"Well," Jim says, "it'll keep until morning." He takes a deep breath of the cold mountain air. "No use worrying about it," he says grandly, "what's done is done." He rises to go to his tent. But finally he can't resist. "We heard you a long way off," he says kindly, as if somehow afraid of hurting George's feelings. "What do you think it is?"

"Shit, I don't know," George says. "I think I tore out its *soul*."

Jim considers this. He closes his lips tightly, then starts to say something, then decides not to. He watches the fire a moment longer and walks off to his tent.

"I was driving with the window down," George tells me. "You know that river bottom where we've been seeing all the elk? Well, they were out there again. They were everywhere — a sea of elk. I was driving along the road, the windows down, just enjoying the road. I was looking out at all the elk in the meadow. I could see them in the moonlight, could see the bulls' antlers, and the cows and calves — and

I was just driving, relaxing. Getting ready to go home, you know. Tomorrow." George glances at his watch. "Today."

I don't really want a beer, but I can't sit here and watch him drink all six. I open one and take a sip. The beer's cold, and the night's cold. The fire is down to dull red ashes.

"Suddenly they were everywhere," George says again. "Elk in the meadow, elk in the road all around me, stampeding. I was driving through their midst — the whole world was elk. There were elk in my headlights. I could smell them. I swerved off the road and knocked something out from underneath Jim's car. That's when it started making that noise." George looks up at the sky. "But it still ran. And I figured the damage had already been done. So I drove on to Chama. I'll bet I woke the whole town. Drove back slowly — the car won't go over ten miles an hour now. Shit," he says, "I fucked up."

"At least you didn't hit any elk," I tell him. But he doesn't even smile. He just lowers his head, braces his arms on his knees.

I could blame George's road mishap on a lack of predators in the ecosystem. I could tell him that the car is but a metaphor for the reality right before our eyes. We have discarded the mountains' wolves and grizzlies, and now the mountain is discarding our machines. I could quote Leopold: "If the land mechanism as a whole is good, then every part is good, whether we understand it or not. If the biota, in the course of aeons, has built something we like but do not understand, then who but a fool would discard seemingly useless parts? To keep every cog and wheel is the first precaution of intelligent tinkering."

"I think I tore out its *soul*," George says.

∗

In the morning everything looks better. That's easy for me to say, as it's not my car lying wounded down in the bushes, but everyone's cheery, and the sky's blue, and while it felt good to be in the mountains, it feels good to be leaving, too. We all have family, in the four corners of the country, and today's the day for dispersal. There's a crisp north wind, and in another week or two, three at the most, the snows will come.

Jim Tolisano is telling George it doesn't matter, it's an old car, and whatever the problem, it never costs more than a hundred dollars. He's being gracious about it, a real gentleman, and George, who has wolfed down a quickly prepared skillet of powdered eggs, is looking better, nourished by Tolisano's cheerfulness.

All through the woods there are the quiet sounds of aluminum tent poles being folded together, nylon tarps being rolled up, sleeping bags being zipped and stuffed. We load everything into the huge rear maw of Doug's truck, the one that used to be Edward Abbey's. I think how Abbey would love the irony, that a hated machine has outlasted him. I like seeing the old truck, though, especially with Doug behind the wheel. It's as if Abbey's spirit still roams the Southwest, not just in the warm winds over slickrock country, but in the presence of four big tires humming down the pavement, a chunk of iron and internal — infernal — combustion hauling friends and family over mountains and across deserts still, the truck going forever on its journey, like an old horse that has lost its rider, but still remembering the places where horse and rider paused together, and lived a life.

Of course, it's just a truck. But it's Abbey's truck, and that's important to Doug, and it's important to me. What was here before us matters. The past, like the present, is sacred. These are real things, things you can count on — unlike the future.

I think about Marty's busted-up car from the year before, and now Jim's. Something about these mountains looks down sternly upon the ways of the twentieth century. The first time I drove through this part of the world, on my way to college in Utah, I got a flat tire about forty miles from the nearest town, at two in the morning. I had an old canvas tarp with me, and I kept stopping and tearing up squares of it, which I shoved into the leak using a ballpoint pen for an awl. Each twisted piece of canvas I shoved into the tire's wound would carry me another ten or fifteen miles before I had to stop and fill up the tire again (I had a little cigarette-lighter air compressor with me). I got out of the mountains and bought a new tire in Monticello, Utah, at daylight, but for a long time it seemed that I would never get out, and perhaps my mistake had been in daring to go into the mountains without first asking.

Because Doug said he would, we drive back thirty miles to Lavinia's ranch — meadowlarks in the fields along the road, and cool mountain air — and he stuffs a big bag of chanterelles into her mailbox along with a scrawled recipe. We can see her big house on the hill, but we don't go up. It's a day for leaving, not for saying hello.

Far behind us, Jim blat-guns along in his gut-wrenched car. He's decided to try and limp to Santa Fe the way it is. It may take him two days to get there, but he wants badly to see his family, whom he's seen only for brief hours at a time over the last couple of months. Doug's eager to get home too. It's been a long summer. Up at daylight and living hard all the way to darkness, then living hard into the night. He hasn't seen his family in almost forty days.

We all stop at Betty Feazel's to say our goodbyes. Afterward, Doug and I will drift down to Tucson, pulling his green dory behind us, and with a smaller boat on top of the truck, what he calls a "dork boat," a twelve-foot skiff he's bringing home for his son, Colin.

Every year, it's the same for Doug. He begins to drift north, up toward Yellowstone and beyond, by April or May as the bears come out of hibernation. He summers all over the West, camping as far north as Alaska. He turns around in late summer and starts slowly south again, ending up at home every fall, down along the Mexican border. The migrating grizzly.

Our goodbyes to Ray, Tolisano, Jimmy Stearnberg, George and Dennis, are rushed and brief, and then they're gone, the sounds of their cars and trucks swallowed quickly by the mountains, leaving only more stillness hanging over Betty's narrow valley.

Doug and I work in silence, lashing everything down tight on the old truck, checking the cinches and the trailer hitch. No one is living at Betty's right now — she's renting an apartment in town while workers build a new house to replace the old one that burned down. There's a feeling of sadness all around that has to do with more than the autumn, and is more than the lingering echo of everyone's quick departure.

Is this how it is for the grizzlies, I wonder, as they go into their dens each fall? Does it become rushed for them at the last moment, as winter moves in? Do they break up suddenly as they head to their separate dens, not to see each other again until the spring?

We string up the chanterelles in the boat, bow to stern and gunwale to gunwale, so that as we tow it through the desert, the sun and wind will dry the last moisture out of the mushrooms, readying them for winter storage. The boat is sea green, and goes well with the orange of the chanterelles. We tie off at least twenty long garlands, which remind me of Christmas. The boat is *festooned*.

It's lonely, this between time, no longer in the mountains but not yet home to family. There's a wall of petroglyphs Doug knows about near Bluff, Utah, which includes an amazing panel that Doug calls the work of "the Wolf Man," because the artist, fifteen hundred years ago, signed his work with huge wolf tracks. There's a chance we may get to that wall before dark, if we don't linger.

> >

PART III

TEETH OF HEAVEN

How could we be capable of forgetting the old myths that stand at the threshold of all mankind, myths of dragons transforming themselves at the last moment into princesses? Perhaps all dragons in our lives are really princesses just waiting to see us just once being beautiful and courageous.

— RAINER MARIA RILKE

It's strange to sit on George Fischer's back porch in Salt Lake City on a bright Saturday morning in June the next year, holding our baby, Mary Katherine. My wife, Elizabeth, is here too, sipping a beer. The sun is brilliant, the sky a fresh-washed blue. This new family makes everything look miraculous.

It was in the spring last year that my grandfather died, and then early into the winter, so did, unbelievably, my young and beautiful mother. This is a story of bears and mountains, and yet everything is connected somehow: my friends and family and the woods.

*

On the dawn drive to the hospital, Elizabeth and I saw a coyote run across the road in front of us. We saw deer and geese and ducks, all going north as we went south, up and over the valley's summit and down into the town where Mary Katherine would be born, the first of my family in five generations to be born outside Texas. A Montana girl.

A bright spring day. We had paused at the summit, and as the contractions were not yet too strong, we got out and I took a picture of

Elizabeth, rimmed against the mountains. She was smiling and holding her stomach: this would be Mary Katherine's last day inside.

The first thing Mary Katherine did outside the womb — after I had clipped the cord and lifted her to Elizabeth's arms — was smile, and then she spread her arms out wide and slowly brought them together to clasp them, without a hitch, her fingers interlocking gracefully. After that, I took off my shirt and held her to me, as Elizabeth had done. She was so small, but so strong. We let her lie like that, here and there, listening to each heart, memorizing things.

When we returned home some days later, the golden afternoon light was reflected by the pond into every room of our cabin. The gold rippled across our faces. We sat down together by the big window and watched the wild ducks and geese in the pond. They called out to one another in busy clucks and honks, and as they did we held each other close, so Mary Katherine could hear our hearts beating along with the sounds made by the birds.

*

Since we were last in the San Juans, Doug and Dennis have managed to get Round River Conservation Studies under way. Students are in the San Juans now, and we're going to head south to meet them. Jim Tolisano is the school's research and field leader, and the kids — there are six of them — are learning everything they can about the ecosystem from him, measuring and evaluating the health of the mountains and getting college credit for it. Tolisano works with them all day in the field, then gives an hour-and-a-half lecture at dusk back in camp.

"If there'd been something like this when I was in school, I'd have done it," George Fischer says. "This is the kind of thing I was looking for." Though George wound up a computer programmer, and one of genius at that, he also obtained degrees in biology and physics.

We've gathered in Salt Lake City for lunch: Italian sausage, chili peppers, eggs, bacon, cheese. Diet be damned today. Dennis and his wife, Trent, are here too, at George and his wife Meike's house. A freelance radio reporter has joined us, Scott Carrier, who is going to make this year's trip.

We talk about the last piece Scott did for National Public Radio, which involved trying to chase down antelope on foot. Scott had heard that certain Indian tribes, like the Tarahamara, would set up relays to chase the antelope and catch them in that manner. Antelope sprint at sixty miles per hour but can't sustain that speed, and that was how the Tarahamara caught them. But Scott couldn't find anyone, white or native, to run with him, except for his brother, who later said that Scott was in "pretty good shape."

"Best damn thing I ever heard on the radio," says Dennis.

"Funny," George says. "You hear a lot of huffing and puffing. They carried the recorder with them as they ran. They ran all over that prairie."

"We might have had a chance," Scott says. "You're supposed to pick one and stay with it, wear it down. But one thing the antelope do is run over a hill when they know they're being singled out. They would get in a herd and hide with the others, and when we came over the hill, the herd would break off in different directions. So we were always chasing fresh antelope. If we could have found a way to stick with just one antelope the whole time . . ."

This sounds like the kind of person we need to look for bears in the San Juans.

I've been in Texas and Mississippi, visiting family for a week down at sea level. I returned home to Montana for two glorious days with Elizabeth and our new daughter before we all headed south to Salt Lake City in the truck. I'll leave Elizabeth and Mary Katherine behind with Trent and Meike when I go to the San Juans. I'm in that thin-air state of being both tired and exhilarated. I'm exhilarated because it's good to be back with old friends and in the company of my daughter — to be holding her on my lap and in my arms.

George is the other kind of tired. Tired of the big city, the long hours, the coffee machine fill-ups, the pressure, the office, the end-of-day fatigue, the computer terminal, the telephones, the traffic. He's listless, like a dog that's been kept in a kennel too long. He's been escaping into the world of the mind as much as he can, trying to read his way out of this work-caused spirit-funk, since he doesn't have the time to blast his way out of it in the way he would prefer, by a deeper immersion in the world of the body. Walking. Camping. Building a fire. Cooking a meal. Sleeping beneath stars.

George has two weeks of vacation this year, and he's going to spend one of them in the San Juans. "I need to get into the country," he says. "I really just need to go sit under the trees and think about things."

And here is the good part: Scott says he likes to drive. He's got an old Saab and a box full of cassettes, and all he wants to do is drive. George and I want to sit back and rest. Dennis won't be making the trip; he's got meetings with wildlife officials concerning Round River's predator-study project in Utah's La Sal Mountains. And Doug's not here either; he's off in Kathmandu or Russia.

*

We go deeper into southern Colorado. It feels as if our hearts are getting larger or as if, the closer we get to the San Juans, the greater the possibility that they have to become larger. Scott is still in love with the view through the windshield — the yellow lines scrolling toward us, disappearing beneath the hood.

The night having finally cooled, we make a late-night stop at a grocery store in Durango. Our car is the only one in the parking lot, under a lamp's bright glow. George stays in the car, resting. Scott goes in for some apples and oranges; I go to the pay phone to check on my little girl, whom I have not seen in over ten hours.

Scott comes out with a bag of food, and I decide two things at that moment: I want a bottle of orange juice and I need to use the restroom. As Scott heads to the car, I hurry into the store to avoid holding us up too much longer.

I buy the juice, then go to the restroom, back between the empty crates and boxes. I always forget to pack toilet paper before a trip, and so a habit I've fallen into is spooling off a modest amount, three or four days' worth, from whatever last outpost of civilization at which I stop. This act of petty larceny is strangely reassuring. It's become, over the years, a marker, a delineation — the final gesture between town and woods.

I finish and hurry out across the parking lot and jump into the car. The first thing George asks is "What did you steal?" He's certain I left without paying for the bottle of orange juice tucked under my arm, so hasty was my exit.

"You look guilty as hell," Scott agrees.

"Just a little toilet paper," I tell them, pulling out the wad of it from my pocket to show them. Then I get myself in deeper. "It's something I always do."

When I tell them this, they howl. So mild is my demeanor that to be a toilet-paper weirdo raises my stock in their eyes.

We arrive at Bruce and Lucy Bailey's and Betty Feazel's ranch and pitch our tents in the cold darkness beneath old-growth ponderosa pines, with owls calling and a dry wind blowing.

*

We breakfast in Pagosa, at the Elkhorn, our gorging ground before and after every trip. We're scouting around for Dave Petersen, a writer who lives in Durango and who's joining in the search.

While waiting for Dave, we saunter into the next-door hardware store. In the store there's an entire stuffed bear on a shelf above our heads. Because it is a small bear, and because it is displayed so prominently, I immediately assume it's one of those blond San Juan black bears. But the closer I look at it, the stranger that bear looks.

It doesn't have the pronounced hump of a grizzly, but maybe the long-ago taxidermist messed up. It has long claws, especially for such a small bear. It could be that after death the skin has pulled back from the claws; still, they're much longer than even an average black bear's. The claws *are* curved, however — not dagger straight, like grizzlies'.

But still, the face is rather round and the muzzle is short, grizzly-like, and the ears are small and pig-like.

George and I circle the bear. He seems to have been frozen in this spot for a long time, perhaps a quarter of a century. Aloud we wonder if he's one of the last of the San Juan grizzlies. We turn over the possibility that he might have been listening in on our many conversations about where grizzlies might be found. But he's a black bear, surely. A very blond phase of a black bear.

We find Dave out on the sidewalk, shake hands, and leave town while it is still early. We'll stash Scott's Saab in some alley and take Dave's old truck up higher into the mountains.

*

We're scheduled to meet Jim Tolisano up at the wilderness boundary. He will linger behind a trailhead sign and wait for us to cross over.

We find the trailhead with the usual amount of difficulty. Tolisano's directions were vague. He'd indicated that the old trailhead sign was broken off at the base, and so we have to look for an earth-colored stub of wood, perhaps already covered over by trillium, fir and spruce needles, and violets.

Back and forth we go along the dirt-and-gravel road, looking for our turnoff. We take a couple of dead ends and then find what seems right, a double-rutted cattle trail beside a rushing creek. Because of the nature of what we may find, I will once more make up the names of places, but I'm reluctant even to give this creek a made-up name. I wish I could just tell the truth. No, strike that; I am telling the truth. What I wish is that I could tell the facts, that this could become a useful part of the historic record of grizzlies in Colorado. And maybe someday there will be so many grizzlies verified or recovered in southern Colorado that I can go back and fill in the proper names. But as long as the grizzlies face extinction, honoring them and their secret places takes precedence over everything else.

We lock Dave's truck, tighten our bootlaces, lift our packs, then mill around, letting the lubrication of the wild seep into our knee joints, our ankles, and between our vertebrae.

Before starting up the trail after the others, something rivets my attention: an aspen, its bark scarred by numbers and letters. What I see doesn't at first sink in. Just another person with a knife, I think, and a yen to have his passing witnessed. "Lloyd Andersen, 1967," the carving reads. Finally recognition stirs: Lloyd Andersen, grizzly killer — the government trapper who in September 1952 killed what was thought (until the Wiseman grizzly of 1979) to be the last grizzly in Colorado, the last grizzly, indeed, in the whole Southwest. Andersen's shadow hangs over the entrance to the rugged canyon before us like a ghost.

I'm fixing to start up the trail with three friends, but I'm hanging back a bit, feeling yet another ghost, a kind of emptiness. Where is Doug? I'm a little discouraged by how subdued the woods feel in his absence. We enter the dark woods, walking alongside the creek. It feels like we're leaving Doug behind, so much is he with the mountains. It doesn't feel right.

But there is work to be done, what Peacock calls "the world's work." It's late morning, but hot already. We walk a ways, then stop, it seems, for a rest. But no, it's not a true rest; it is a calamity of the bowels, that good Monticello, Utah, cooking we had on our way south, now coming back to hamstring both native sons, Scott and George, who'd ordered get this — the spaghetti plate, a scant twelve hours ago.

Dave and I share a cigarette, as does Scott upon his return, less afflicted. When George returns he's a bit pale, but heroic-looking. We rest a bit longer, like soldiers. We pass a canteen of water around and then get up and walk another fifty yards, around a bend in the trail, to where Tolisano has been waiting for us, down by the rush of the cobblestoned creek.

Time for another smoke, a bit of lunch, and another quick trip into the woods for George, who's clutching his stomach. Jim's truly concerned and sympathetic, a generous response given that we've all just finished belly-laughing through the umpteenth retelling of how George, on his midnight beer run last year, mangled Jim's car.

"Whatever happened to that car?" George asks on his way into the woods.

Jim shrugs. "Totaled," he says cheerfully. "Would've cost more to fix it than it was worth."

George looks stunned with the responsibility.

"It was old," Jim says. "It's okay."

"You're sure?" George asks. He's a big man and he'll do what's right. Buy a new car if that's what it takes.

"I'm sure. If you hadn't run it off the road, it probably wouldn't have lasted another . . ." Tolisano shrugs again. "It was just old."

We wait for one of the Round River students, Steve, whom Jim has sent down the creek and out to the road to look for us. We begin to feel prickles of worry over Steve, and then we see him walking back up the creek. Peach-faced, Steve is young and very tall and wears an old white canvas tennis hat. His shirttail is untucked. He has stubble on his face, a couple of weeks' worth, which indicates how long Tolisano and the students have been in the woods.

So far, we're averaging about point-six miles per hour. And what I have previously been calling a nameless creek — that won't do at all, now that Tolisano has us in his clutches. From here on out it's Rio Diablo.

Tolisano starts up the trail with the verve of a man on a roller coaster. "I'm a little worried," he says. "It's the first time I've left them on their own, and I'd like to get back before dark." But his students are secondary reasons for the electric pace. The main reason is that Tolisano simply likes to run with a pack on his back.

After a brisk trot we find ourselves standing at Rio Diablo's edge. Most western rivers and braided streams trade depth for width and speed, but the Diablo seems to be all of these things and more: cold, wide, deep, furious. We cast up and down the rocky banks for a fallen tree. We finally find one suitably precarious for Jim's interest and use its mossy, rotting, branchy spine to cross the Diablo.

We scramble up a steep chalky cliff, hanging on to exposed roots, climbing hand over hand with loaded packs toward blue sky, never pausing long enough to look down and acknowledge how tiny the creek is fast becoming. Dave's ahead of me and Tolisano's above us all, from time to time peering marmot-like from over a ridge to evaluate our progress. Now Dave slips and goes down hard, starts to slide, grabs a tree, is saved. He gets up and continues on, and when I come to the spot where he went down, there's blood in the sand, a long bright smear of it. This is a journey Peacock would be proud of: into the haunt of the bear, leaving behind, in Peacock's words, "tatters of flesh hanging from the bushes" as if in offering.

We gain a view up into an old-growth Douglas fir forest, several hundred feet above Rio Diablo. "The country gets pretty rugged up ahead," Tolisano says, staring at the pyramid-shaped mountain on the horizon, at whose base the camp is pitched. "If we didn't climb up here, we would've gotten into a bunch of steep cliffs," he says, and peering through the trees, I see what he means by steep: vertical gray dolomite, hundreds of feet of it, all towering above increasingly savage rapids. "This is some of the most rugged country in the San Juans." Tolisano points up toward the Diablo's headwaters. "Great security. A bear could hang out up there and never be seen. We've been meaning to check out this drainage for a long time."

What's challenging about the route we're taking is that although it parallels the rushing creek, the terrain — brush and downed timber, mostly — is punctuated by a never-ending series of steep ravines. The result is that we clamber along on the steep side hills, crooking our feet and knees at ridiculous angles, thinking all the while, *This is the worst kind of hiking.* And just as we think that, we see Jim's back, far ahead of us, drop down into one of those hilly ravines, all of them sporting waterfalls. We must slide down to the bottom each time, sacrificing all the elevation we'd fought to gain, and then go straight up and out of the ravine, earning it all back.

The temperature differentials are startling. When we pass from shade to sun, we go from icehouse cool to shimmering heat. I've read that in the Rockies the temperature can fluctuate by as much as twenty degrees, and the differences in these woods seem to be at least that. There's no doubt where the bears will be: panting, grunting, heavy-caped furry vessels, they'll sleep during the day in the deepest, coolest shade they can find — near a spring, perhaps — and come out only at night. They'll be up in the deep old forest, and if they happen to flutter their eyes in sleep and look out at the bright hot sky far above, perhaps they'll think, in groggy dream-sleep, that summer and its heat are only their imagination.

Perhaps I'm confusing what the bear might think with what we're thinking as we scramble up and down the side canyons, moving slowly but steadily up into the Diablo's jungle as Jim thunders on, racing for camp. Inexorably — the wave of us moving through the woods, sometimes following game trails — we draw closer to the huge gray pyramid mass that takes up, as we get closer, all of the northern sky.

We inch along, passing beautiful peaks in the shadow of the Continental Divide. Cool breezes slide through the trees despite the June heat, which threatens to explode bare rocks in the sunlight. We creep past one faraway peak, a cone-shaped mountain with a short protruding pinnacle, called Nipple Peak. It is so aptly named that it causes a stirring in the heart and the groin, the size of it so immense and the shape of it so perfect.

Walking, forever walking. Tolisano has mastered the fastest walker's nefarious ploy of stopping far ahead — resting, sylph-like, on a mossy boulder at the crest of a steep pull — then leaping up again just as the stragglers arrive, gasping, sucking, vacuuming up air.

Lightning bolts still split my vision at odd times, but there is no pain. It's just a mystery. It's like a secret, a thing that only I can see.

W e near the confluence of Soldier Creek, where it pours down out of its glacier with thunderous vigor into the waters of Rio Diablo. That sound! We stand on a dolomite cliff looking across the side gorge of Soldier Creek. Round River camp is across that gorge. It's early evening and the sun is down behind the high mountains. Cool light is

flowing through the canyon. We pick our way down a vertical crease in the cliff.

At the cliff's bottom we find ourselves in another old-growth forest of fir and spruce, trees two and three hundred years old. Man has been here before us, however — man and helicopter — a hunters' semipermanent camp, long since abandoned. Picnic tables, lawn furniture, dutch ovens, and ragged, rotting plastic are scattered in all directions, as are dozens of rusting tin cans that seem to have fallen from the moon. We stand for a while in the camp's squalid center, ankle-deep in refuse. Propped against a giant centuries-old cedar tree is a single sheet of plywood the size of a large picnic table, a billboard with the words "Odegard Well Logging, Durango."

To depart from the facts at hand, which are, simply, *trash in the wilderness,* would be judgmental and risky, but it's hard not to believe that the old boys came in with their oil company's helicopter and set up the realm of their private lives, their private October lives. By November, of course, they would have turned their attentions elsewhere. Lake Tahoe, perhaps. Christmas in Vegas.

"I can report them. I can find out which outfitter uses this district," Dave says, without much conviction, as we realize that those honyocks are not likely to be the kind who apply for permits. That, and the fact that the damage has already been done.

"They should have someone come in and clean this up," Dave says. "I'll tell someone about this. I'll tell the district ranger."

It's a good idea. Instead of below-cost, tax-subsidized logging and road building, workers could clean up the forests, not level them. It's not a new idea. Conservationists have been talking about it ever since the New Deal. The footbridges that the CCC built across creeks to prevent erosion have held up well.

Such are the pipe dreams of environmentalists: a workforce in the woods that respects the forest.

We leave the pigsty camp, moving through more old growth along Soldier Creek, searching for a place to make our last crossing. The pitch

of the creek grows quickly steeper as we move toward its glacier. We finally find a spot to cross where the water is knee-deep and fast between giant jagged rocks. We sit down on the cobbled beach — lush ferns in our faces, and ever the fabled Heracleum, bear food — and slip off our boots and socks, bundle them in one hand, rise and ease our hot feet into the cold stream. We unbuckle our pack straps in case we slip and go down. (It's likewise tempting to drape our boots around our necks or over our shoulders, but in a fall they too could choke or drown us, so we carry them free and clear, like live grenades, in our hands.)

Across the stream there is one more cliff, strewn with loose talus and yellow dirt — a mineral seam, a fracture line — another chimney. Camp is in the big woods above us, three hundred feet up. A raggedy yellow rope of frayed nylon, the kind of rope you buy in a dime store with the spare change in your pocket, hangs limp from the top, appearing no more substantial than a spider's single strand. When I grip the fraying rope, which is about the diameter of a pencil, it stretches like taffy and feels rotted.

Jim, of course, has scooted straight up like a monkey. He crouches on a dolomite pinnacle far above, smiling. In the quick-descending gloom, with Jim perched on the promontory like that, his bright teeth look more like the grimace of some animal, a guardian waiting to consume the few who might somehow make it all the way up to his lair.

Once I give in to the notion of possible death or disfigurement, it's kind of fun, ropewalking up the cliff, taunting my heavy pack, which threatens to pull me to the rocks below. In a world full of poisons and murderers, at least if I fall it'll be my own damn fault — I chose to put too much weight on a rotty rope, and that was that.

Jim disappears into the woods just as I pull myself to the crest. I follow a winding game trail toward the smell of a camp's woodsmoke.

I catch up with Jim at camp's edge. He's pausing back in the trees, as he did when we found him lingering at the wilderness boundary earlier in the day. He's watching his students move around by the fire,

preparing supper without him, surviving in the wilderness unaided. Why, they're children, I tell myself.

We walk into camp together. It bodes well for the future of the San Juans to see the absolute joy with which they receive Tolisano back into camp — all but leaping to their feet and yipping like young wolves, with the alpha having come back bearing game, or in this case three journalists.

Their eyes are so bright! I see a lot of young people in schools who claim to want to study and write about and even save the woods, but too many of them are partyin', hang-glidin', do-nothing Lycra yuppies. But these children of Round River, these scruffy kids — right away you know that you have stumbled into a camp of warriors.

Introductions are made: Chad, Marty, Dan, Beth, Sarah, and Tolisano's assistant, Jim Sharman, a young Peace Corps worker who is between jobs. Almost all of these kids are new to the Rockies. Beth takes me to "the shrine" before the last light fades. Tolisano picked this camping spot from a topo map, but no one knew how beautiful it would be until they arrived: a long lush meadow on a shelf above the confluence of Rio Diablo and Soldier Creek, with the mountain looming above us; a rocky, forested spine (with swamps, bogs, and springs — at nine thousand feet!) running north upriver along Rio Diablo's cliffs. A harbor. The meadow is full of purple-and-yellow irises. Fog rises from the meadow, and around the perimeter, respectfully set apart from one another, are the students' tents.

The shrine is centered around a pile of bear shit that was sitting at meadow's edge with a same-day freshness when Tolisano and the students first hiked in. The pile was right where they wanted to camp. It was a good-sized shit — not mature-grizzly-sized but probably from a full-grown black bear. One thing these students (and Tolisano, Peacock, and Sizemore) are keen on is omens, so they set up camp here. Over time, they have brought various woods treasures in to place in a circle around the good-luck shit. A mouse's skull. A bundle of irises. A raven's feather. An old blue-speckled coffeepot. A hunter's aluminum arrow, snapped in half, as if in an offering of peace. More bundles of flowers. At the edge of the shrine there's a black plastic tarp on which nearly two dozen bear-scat samples are drying. What must any bears passing by in the night think, coming upon such a tarp?

Beth kneels by the shrine and lists the names of the flowers bundled there. Her naming of them seems to drop her into a dreamy state. I realize she is no longer talking to me, nor even to herself, but to the flowers and to the mountain. I leave her kneeling there and drift back to the fire. George, Scott, Dave, and Steve have ascended the rope cliff and join us, still sweaty and breathing hard.

The sun is long gone, but the twilight glow of the earth's cooling illuminates us. Hiking into camp, I saw a small pile of ashes in the woods where one of the Round River students had cleaned out the remains of the campfire. The ashes glowed fluorescent blue in the dusk, like the keratin of fossilized fish scales, but these ashes were from sticks and branches that only a year or two ago had been alive. In all my years of camping I have never seen blue ashes, have never seen such iridescence. It is a small mystery that remains unexplained, unsolved.

*

I pitch my tent, as I have so many times, in the dark. I make a tiny five-stick fire and heat a bit of water for the foul and dreaded back-packer's special, Top Ramen. I like being alone, away from the other fire, the other campers. Alone in the woods by the iris field. Not all the way into the heart of the San Juans, but perhaps cradled in the crook of her arm.

The East has its Adirondack National Park, a complete bioreserve. And the San Juans, think of it! An entire web of life preserved, stretching from the arid plains into the foothills, up the rivers into the mountains, the glaciers, the heavens — a place where all that exists under God's eye can be allowed to play itself out naturally, and where we can study and observe what the world was once like before we began to carve on it, before we began to make mistakes.

My water begins to simmer. Grand dreams! We don't want to turn the whole West into one big bioreserve, one entire region full of peace and harmony and a complex, healthy web of life, for God's sake. Maybe those dreams will follow — but for now, just one place to start with — the San Juans — one grand project. All generosity is an act of courage, says Barry Lopez.

It's dark, a new moon, with lots of stars. If it were a full moon, I might be able to look down on this iris meadow and see a grizzly feeding in the moonlight, ripping and plowing the tender grasses and the lush earth, stirring up the scent of wild ripped onions.

I extinguish my fire and watch the stars. This is my first camping trip since the birth of Mary Katherine. It feels good to know of a few wild places that I can show her. A few magical spots such as this meadow, this fast creek in the shadow of a glacier.

How can there not be woods — wild woods — for my daughter to move through? I cannot conceive of such an earth.

I go down to join the students. The pitched tents around the meadow's edge give the impression of a hunting camp from a hundred and fifty years ago — or if not hunters, perhaps even the hunted, having re-treated so far into the mountains. This place, like the Hoodoos of last year, has an aura of power around it, and it doesn't take extraordinary sensitivity to feel it. It's real. It's magical, inspiring, comforting.

We talk around the campfire about the only San Juans bear yet seen by Citizens' Search, a big brown-colored bear that George and Marty saw in June of last year. They saw the bear twice, in quick glimpses, too quick to be certain that it was either a grizzly or a brown-colored black bear. The second time, the bear came toward them, trying to check *them* out, before it turned and ran into the brush. The significant, thrilling aspect is that the bear was bigger than Big George.

Around the fire, Dave Petersen, a hunter, discourses about the difference between good hunters and slob hunters. He tells the story about how bear hunters will dump slop buckets of leftover doughnuts and old meat in the same place in the woods for several days or weeks, to lure bears and get them used to coming to the "dump," where, once the season starts, the "hunter" can shoot them.

"So that's what those empty buckets were," George says, talking about some trash he found in the woods last year.

Glen Eyre, the veteran Colorado Game and Fish officer, believes there may be as many as a dozen grizzlies in the San Juans, and he's in favor of closing black bear season here immediately. It's refreshing to know that there are still officials these days who attempt to base their wildlife management on biology, not politics. Whether Eyre's recommendation will be accepted by his superiors remains to be seen.

That's the genius of Round River — the genius of Dennis Sizemore, of his being a good listener. For thirty years now, biologists have been leaning hard on the pursuit of animals wearing radio collars. But Dennis wants to look for tracks, for scat, for hair, and to sniff the breeze. He wants to listen to the land's citizens, not just the bears and coyotes and elk, but the people too.

Any existing local tolerance must be discovered and nurtured. Any absence of tolerance for grizzlies must be acknowledged and dealt with, not ignored. Our job — Round River's — is to educate and give harbor, to protect wildness.

It was from the loss of wildness, the loss of vigor, that all the rest of our country's bleeding social ills accelerated. It will take generations to fix these ills, but if we can help change attitudes, we might be able to get there just ahead of the train wreck — might be able to protect the last roadless areas and begin to repair those places that have been harmed. And if we can help change attitudes toward the land, all the other wrongs of domination (rather than caretaking) in society might reveal a pattern, a model to follow, so we can turn things around.

Our side. There is no other side.

The Colorado Division of Wildlife still hasn't acted officially to acknowledge either the presence or even the possibility of grizzlies, but we can sense a turning despite the denials. A recent article by

Todd Malmsbury, the chief of public information for the CDOW, prompted a new outpouring of reports of rumored sightings, including the sighting of a northern New Mexico grizzly by one anonymous letter writer:

> . . . I read with interest your article in *Colorado Outdoors* concerning Grizzly Bears in Colorado. I was tempted to write this letter when the incident concerning the outfitter killing the grizzly . . . was reported in 1979.
>
> I had an incident of my own concerning a grizzly that may be of interest to you. It was about 1965 when I owned a fairly large sawmilling operation in . . . N.M. Since I had been receiving logs from [a certain company], their manager, [Leo], came by my office and suggested we take a ride up one of his main logging roads to see if the snow drifts had melted out. It was April and a sunny nice day so we went up in a four wheel drive Suburban. Before we reached the snow line we noticed, through some aspen trees, that there was a light brown animal of some kind across a small draw in a clearing. Leo said, "Looks like there's a bull elk in that clearing." By then we had coasted to an opening in the trees and saw to our great surprise a very large light brown bear in the clearing facing us. We were stopped and the bear had not seen or heard us. He was about fifty yards away on this grassy sunny slope eating grass. He was quite large with a huge head and front legs that were extremely bowed with his toes pointing way inward. We talked about his beautiful coat that was rippling in the breeze. After ten or fifteen minutes we could tell that he was getting wary and uncomfortable. He started swinging his head from side to side probably to test the wind. After an extended period of time he gradually turned and walked forty or fifty feet into scrub conifers and aspen.
>
> We both felt extremely fortunate and mildly excited about what neither of us had ever witnessed before. We were both very experienced in the mountains and wildlife in general. We could not figure out what kind of bear we had just watched. He was not like any bear we had ever seen. He was shaped much different in all ways and his fur

seemed much longer than any black or brown bear we had ever seen. We mentioned he looked like a grizzly but tossed that thought off since we knew grizzlies were extinct in all southern areas. The plot of this bear story did thicken.

After the bear disappeared we continued a number of miles up this road to where the snow did block the road. As we came back down the road I suggested to Leo that we stop and see if we could get a measurement of the bear's foot since his cowboys had been reporting "tracks" of a very large bear for several years. They said they were about a foot long.

We did stop at the base of the clearing that sloped upward to the north. I had a strange feeling that I do not want to call fear since we did not have a gun of any kind. We started climbing the slope and walked around the head of a small draw to get to where the bear entered the timber. Before we got to that spot we passed under a very large spruce tree that had the hide and bones of a cow from at least the season before. There were many bear tracks there but I couldn't find a complete track. The bear had been moving the bones and pieces of hide of the cow carcass. There was nothing there to eat so I assumed he may have been rolling on the flattened carcass. Leo proceeded into a small aspen grove further up the slope. I went to the left of the spruce tree and was finding some better tracks as I walked up the slope. I heard Leo to my right as he walked through the aspen grove. I had my head down looking for a better track. I became aware of a slight noise straight up the hill from me in the edge of the timber. I knew what it was before I raised my head. The bear was about fifty feet in front of me and walking straight at me. The massive head and bowed legs striding at me caused me to run back down and around the spruce tree after I shouted to Leo that "he's here." Leo immediately came busting out of the aspen grove and flew right by me since I had stopped to wait for him. Leo's eyes were bulging as he went by me. I took out down the hill thinking I was the one that was going to get caught. Leo went around the head of the draw and I cleared it in one leap to beat him to our vehicle. I instinctively went over the top, not wanting to stop and open the nearest door.

I looked up the hill for the first time since I had seen the bear. Leo had arrived and we saw that the bear had stopped under the spruce tree on top of the remains of the cow. It appeared that he was not going to leave so we watched him for quite a while. He just stood there swinging his head back and forth. I do not feel ashamed for running though I hear it is the worst thing you can do. It would take a better man than I am to just stand there or go into an upright fetal position. We finally went down the hill. The bear was obviously guarding the cow carcass even though there was nothing of it to eat. . . .

I should say that the creek we were on the rim of is totally impregnable by man except perhaps on hands and knees and I wouldn't bet on that. There are so many areas that hunters do not get into that bears have no trouble evading them. There are many bears . . . but hunters never see one. I have only seen them over dead animals at dark.

✳

Around the campfire that night, Tolisano tells us what Jim Halfpenny, a biologist, told him. "He swears wolverines are in the mountains," says Tolisano. "Says last winter he followed the tracks of one across three feet of fresh snow, until he lost it to new falling snow." There's a brief fire-staring silence as we consider that knowledge. Halfpenny is not a rumor-monger; his book *Mammal Tracking* remains the definitive text on the subject. Concerning the wolverine — carcajou, skunk bear, the "Blind Swallowing Thing" — *Wildlife in Peril* is summoned: "Even in midsummer it can snow on top. It is a place . . . where there are two seasons: The Fourth of July and winter, a place in which the wolverine, 'The Giant Weasel,' as Seton called him, is supremely adapted to live."

✳

In the morning George hurts. His stomach is raw, his head is splitting, his very soul feels ripped as if by talons. He lies flat on his back all morning, absorbing meadow smells and birdsong in an attempt to heal himself. Perhaps he's suffering from computer withdrawal. It hurts even to look at his head, which is sticking out of his tent, skygazing, unblinking. He's trying to get back to the land of the living, the Big George we know and love.

We can feel so many of the spider-and-prey's tugs upon the web of life. When one of us becomes diminished, we are all diminished. Those of us at the top of the pyramid — bears and wolves and men — are by nature a little brutish, and humans often don't feel the muted strugglings that engender small holes in the web, the gasps and cries at the disrepair caused by our own movements. But in the San Juans and other wild places we can feel more, know more. In the woods we can learn, or relearn, to feel if not understand the movements of smaller individuals. All of our senses can become keener.

My feet are still numb. Some days, for seconds at a time, I'm nearly blind with blue light. But then it passes and I see things more sharply than ever. I find myself thinking once again about the birth of my daughter.

*

The students drift off in pairs. Jim Tolisano has taught them orientation and survival skills, but still he wants them to return from their day's rangings by five, in time for his lecture — and in time to send out a search party in case of trouble.

Dave and I decide to take a short hike farther up Rio Diablo, to check things out. We cross trickling side creeks at the base of cliffs ringed with Heracleum swamps, little mud bogs. The southwestern exposure is a bit hot, and we don't find any tracks.

Dave spots a log dig at the edge of a clearing and measures it — a foot by a foot and a half by two feet, surely a black bear's — and we climb higher, into the fluttering aspen. We pause by a tiny creek and detect a wet-dog smell, probably that of a coyote. Farther up is the sweet warm barnyard smell of elk, and then we find where the elk have bedded down; perhaps our approach moved them off. We examine aspen trees for claw marks, and find several. We sit on a point and look at the other side of the Diablo. Heracleum-green avalanche chutes plunge over the edges of stark dolomite and trickle hundreds of feet down into small talus fields, where Douglas firs give shadow.

At midday, two cow elk gallop across a meadow, fat and healthy as prize horses, their sides gleaming yellow in the sun. They vanish into the old forest. The fungal scent of mushrooms is all around us, rich and raw like sex, though the chanterelles won't be up for several months. As the sun heats the matting of leaves and needles on the forest floor, I imagine that I can smell spores trying to rearrange themselves and take growth beneath us, but it is only the earth, the dirt that will give rise to chanterelles.

There's been a long stretch without any planes. The last ones passed overhead this morning at breakfast, their presence a foul interruption, and now another one drifts by, glinting silver and sounding very loud. Dave and I go higher, out of the aspen onto bare rubble, the cliffs and rock knobs of the mountains. It's not the best place for bear sign — they prefer earth to rock — but the cliffs allow us to orient ourselves in this hidden little valley.

The glacier is dirty white and massive, a wall of protection for the bears. Dave climbs one solitary knob on a ridge and gazes down a two-hundred-foot pitch of talus to a slender fast creek, which is nearly obscured by fir and aspen. He looks dramatic up there on his eagle knob, perched in the blue sky with the glacier before him, scanning the valley. He's got his camera, a professional-looking machine, its crystalline lens glinting in the sun. I ask if he'd like me to take a picture of him like that, all noble and rugged.

There is nothing in the air I can sense to indicate that this is a bad idea. But quick as a spark, just after Dave says "Sure" and begins leaning down from his perch, a strange thing happens: the camera, black and shiny and heavy, slips from his hands. I'm still a good ten feet away. I make a futile lunge for it, and the six-hundred-dollar box falls into a chimney flume of scree far below, and from there the camera bounces from rock to rock, cartwheeling out of sight.

There's nothing to be done, nothing to say. Dave remains standing on his perch, looking eerily alone. The stillness all around us is lucid:

every rock, every tree, every pebble seems outsized and overly specific as we try to tone down our desperate camera-searching eyes to human size, try to corral our wild-hammering hearts and sick feeling. I wonder if perhaps the camera's not still falling, still ripping through the forest in an unstoppable descent.

Dave looks down the chimney and, I can tell, plays it all back in his mind. Then he gives his pronouncement. "Shit happens," he says.

*

We wait for Scott Carrier and the student Jim Sharman to come back before starting class. Finally they appear from out of the dusk to stand, weary, at campfire's edge. We're all gathered around, even George, who has gotten sicker, as if the mountains are shaking him, trying to scold him for having consorted so freely with computers since the mountains last saw him.

We are all "gone" from the mountains a bit more each day we're away, I think, but George's fear, it seems, is that he has gone too far away, and bravely he has decided to take his medicine, to return to the mountains and lie there on the ground and ache and be sick until it is over and he is back with the mountains again. All day long George has talked as if delirious, not about the wilderness but about computers, about hardware and software — getting it out of his system.

Tolisano tells us about MVPs, which refers not to most valuable players, he explains, but to minimum viable populations, otherwise known as the Rule of One Hundred. He tells us that this can be defined as "the smallest an isolated population can be and still have a ninety-nine percent chance of surviving for a hundred years." A hundred years! How our standards have fallen! In earlier periods — the Cretaceous, the Jurassic, the Triassic — species hung around for hundreds of thousands of years. A hundred years is a wan, human yardstick.

There are biologists who believe that in order for a species of large carnivore to survive, to maintain the genetic diversity necessary to have a ninety-five percent chance of survival, a population needs fifty animals. Wolf or bear, lion or tiger — fifty. It is a number that fits into computers and equations easily, and it is an easy number to remember. Less than fifty? Good as extinct, say some biologists. The extractive industry folks say the same thing, that fifty equals zero, and also that extinction has always been a part of the earth.

What goes unheard, or unheeded, is the fact that though extinction is natural, the current pace of extinction is most unnatural, occurring a thousand times faster than at any other period in the earth's four-billion-year history, unrivaled even by the pace of the great Cretaceous die-off of the dinosaurs. Everyone around the campfire must believe that if the bear and the wolf and the wolverine go over the cliff of extinction, then we too will follow shortly, being at the top of the food chain. Some days I feel one more golf course will doom us.

*

"They're real serious about the Rule of Fifty," Tolisano says. "This is a big aspect of conservation biology. They talk a lot about genetic bottlenecks, about loss of chromosomal heterozygosity, the ability of a cell to choose one path or another. Strength — possibility."

Does anyone test cell and chromosome strength against mutation, I ask, or do these academics hold that all cells are identical, all species are identical, all chromosomes and in fact all individuals are equal? Tolisano just shrugs and smiles. He doesn't like the academics either, but he's not the type to waste time getting wound up about it. He'll do his work and let them do their work, seems to be his attitude, and let the best man win, let the species be saved. Good will prevail, though not without struggle. Conserve energy and don't fight these dumb bastards, Tolisano's smile seems to say, but be aware of what they're up to.

I want to know whether biologists tie this Rule of Fifty to the various reproductive ratios and the ability of some predators — like some prey — to recover quickly. Wolves can have twelve, even twenty pups a year per pack if both the alpha and beta females breed. Grizzlies, on the other hand, can have only two or three cubs every third or fourth year, and only then from females that are at least six years old.

"Nope," says Tolisano, "they don't factor that in. They just come up with this number fifty. They call it random broadcasting of genetic drift."

"Sounds like random bullshit to me," says Beth, one of the students.

The indispensable John Murray writes, in *Wildlife in Peril,* that the scientists don't always agree on the Rule of Fifty. Some call it two hundred, some call it one hundred. For long-term genetic diversity, they call it five hundred individuals. In short, they're all over the place, these biologists — everywhere except in the woods.

One scientist (a brave one) has suggested, writes Murray, that "because carnivores are typically homozygous" — that is, not so dependent upon issues of heterozygosity and genetic mixing — "genetic problems may be of less concern to their survival than other environmental factors." Finding ways to say yes or even maybe, instead of "no." What other choice is there in this life?

Tolisano's voice is steady and low, calming. It could be fifty or sixty years ago, back in the days of Aldo Leopold. Knowledge is being passed on around the campfire in a way that is almost like storytelling. But will any of it get farther out into the world? Will any of it be used by people in power, by leaders, or will it all stay a secret, almost mythical?

In the natural world, Tolisano says, there are different "types" of species, and it's a good lesson in alertness and awareness to consider what kind of animal each species is, how it fits into the natural system. It's helpful to consider the various plants and animals in the woods as being keystone species, indicator species, flagship species, or recovery species.

The way each species fits into the whole of the system, says Tolisano, the manner in which it contributes to, and is sustained by, the system — that's the proper way to assess diversity in a system, and to defend its needs. "There's too much linear thinking going on," he says, "too much management for a single species at a time." Relationships must be addressed. Not just the grizzly bear, or the elk, but everything around them.

"If you take out a keystone species," Tolisano says, "you break down a lot of the key relationships in an ecosystem. A keystone species plays a large part in the presence or absence of other species." Both predators and prey can be keystone species: the barren ground caribou of Alaska, for instance, upon which both wolves and grizzlies feed at different times of the year. "An indicator species," he tells us, "is a species that is very sensitive to change. Often the slight increase or decrease of a soil nutrient can affect an indicator species. Many plants are indicator species." It crosses our minds that what we're looking for, the grizzly, is a classic indicator species.

"There is a third species used for measuring diversity," Tolisano says with a slightly embarrassed look. "A flagship species, the one that gets people jazzed. It makes them pay attention." He shrugs. "A sexy megaphone."

"The terrible flat-faced bear," George mutters — his phrase for the human species. Yes, George is still with us, sitting upright but shrouded in his sleeping bag.

Tolisano nods. "Often a species will overlap, will fit into two or more of these classifications." Although what we are doing up here is most definitely unsexy — picking up shit in the woods — it occurs to us that the grizzly would also have to be called a flagship species of the San Juans, or of any other region where it exists. I remember a conversation I had with Doug Chadwick, a writer and biologist who lives on the north fork of the Flathead River in Montana.

Chadwick was enthused about a biologist he'd met whose specialty was beetles, and who was lamenting the fact that flagship species always seem to be the driving force in wildlife management policy and public awareness. "This guy said we have it all wrong," said Chadwick, "that if we think it looks bad for the wolf and the grizzly, we should come down to his level and see how dire it is for beetles, and for the microorganisms — *that's* the most accurate way to measure the health of the forest." Chadwick spoke with the awe that's roused by a new way of seeing the world. *"Beetles,"* Chadwick said, elongating the word.

"Flagship, indicator, and keystone species are a great tool," Tolisano was saying. "We can't understand every possible relationship in the universe at every level. We're physically incapable of holding that much data, even if we could discover it. Our minds would explode. So in the meantime, we're looking for something we can follow — he smiles, the phrase sounding eerily spiritual — "something that will tell us about the quality of life here. We're looking for something sensitive to change, and it's best for this kind of work if the species you're trying to study has a large areal requirement. That way, your conclusions will have merit, in the way that a poll relying on a sampling of ten thousand people is probably more accurate than a poll of a few dozen people."

The notion of "large areal requirement" for studying a single species reminds Tolisano again of the dreaded Rule of One Hundred and of minimum viable populations. "Wolverines," he says, "never have large populations to begin with." The Rule of One Hundred and the Rule of Fifty are abstract, dangerous, the enemies of hope.

There are only ten of us seated around the campfire tonight. Would we give up, would we want the appropriate management agency — in our case, God — to sign off if our numbers, in the San Juans or in Yaak, or in DeKalb, Illinois, fell below fifty? Would we be so blithe about relinquishing the reality of what Bill McKibben has called "this teeming paradise"? I think not. But if we sign off on the bear, we must prepare to have the same done to us. This is the law of physics. Things set in motion tend to continue in that motion. The sound, all around us, is that of falling dominoes.

Tolisano goes on, as if to drown out the sound. "There is a fourth kind of species to consider, and that's a recovery species, which can indicate that the system's coming back to life, recovering from the ground up, which is usually the first place you see recovery. Soil, flora and fauna, and the little stuff you find in water — rotifers, dinoflagellates, and the like — are good recovery species. As, again, would be the San Juan grizzly. It would indicate a recovery of respect and tolerance. It would indicate — as the grizzly does so well, wherever it lives — the presence, or recovery, of some small state of wildness."

Tolisano stops for a moment and looks up at the stars. "Wolverines are particularly susceptible to a loss of wildness," he says. "As the pristine country disappears, so do they. They don't adapt well to civilization. The same is true for grizz."

A shooting star crackles past us, directly overhead, as if bound for some important destination.

*

The night lengthens. Tolisano talks for a while about the moral decision of not using bait to find bears; likewise, we will not use radio collars. Instead, Dennis, Doug, and he want Round River students to consider how to think like a bear, and in so doing they will come to a better understanding of bears and of the whole of the woods.

"I want us to become more intuitive," he says. "We're going to incorporate local and traditional knowledge into our search," by asking the nearby residents where they think the bears are. "We're going to do inventory work, biological assessments." Baseline data: *this is what we have here, this is what we want to keep.*

Tolisano recalls a trip he made to Sri Lanka, where he tried to inventory and save some wild country. He'd heard of a man there, a Professor Subreimenov, who knew the name of every plant, every tree in the forest. He knew everything about the soil, the water, the animals — but especially about the plants. For years he'd been the only man who knew the name for everything, and when the villagers had a question about a plant or its properties, they'd ask Professor Subreimenov.

"He died, finally," Tolisano says. "He was an old man, and all the natives were heartbroken. The man who knew more about the botany of that country than anyone else was gone. Everyone just sort of wandered around in shock. All that knowledge, gone."

Where did Subreimenov's knowledge of the land go? Back to where it came from, into the earth.

I n the morning, I decide to head into the high country Dave Petersen and I looked at yesterday. We'll hike into the old-growth
Douglas fir right below the tree line, then perhaps up and over the
mountains and across the snow. George is still down and out; maybe he
should be in a hospital, but this is where he wants to be. Scott is talking
with Tolisano and a few students, taking notes.

Like so many of us, Scott's a tad goofy. He sometimes seems to spend
a half beat too long in dreamland, which might be the result of all the
radio interviews he's conducted. Sometimes his conversation spills out
too quickly, his thoughts leaping two beats ahead, like a frightened frog
jumping into a marsh, and you have to hurry to catch up with his
conversation. From the fragments I hear, Scott is asking the students
about Doug Peacock, not bears.

I stand at the edge of our meadow and look at all the bear scat the
students have collected. I remember how, that first summer, Doug and
Marty and I, full of pep, covered so much ground but found no scat up
high — too close to the sheep grazing, it seems. Sometimes I feel that
we might as well be looking for the yeti: footprints, a hair here and
there, a sighting, a blurry photograph. How far into the woods do we
have to go?

Tolisano comes up beside me and tells me about a friend of his, a
Jicarilla Apache, in New Mexico, who was surprised to learn from him
that there were still grizzlies in Colorado. She paid close attention as he
told her the story, and wanted to know everything he'd found out. He
asked her in turn what the presence of grizzlies in the San Juans would
mean to the Jicarilla Apaches, so far away — whether it would be a
good thing or a bad thing. She said to Tolisano, "It's not a matter of
good or bad. It would just be of great interest to us. It would be a matter
of great power, not an issue of good or bad."

I think of the simplicity of that logic and the clear articulation of it:
for the San Juans to have grizzlies still would mean power for this
country. I'd like to believe that there will always be one more "last" San
Juan grizzly. That one more will always climb out of the earth, out of

the center of the country, just when the power of its presence is needed: a wellspring of power, a small trickle, but unceasing.

✻

I like walking alone. It is as different from walking with a friend as lifting boulders is different from, say, lifting weights. You think about different things. Your rhythms, and the rhythms of the day, are different. Walking alone makes me feel "out there," detached. Sometimes I prefer it, and have become addicted to the lovely way a day stretches out when you have it all to yourself. You can climb the steepest hills at your own pace, without worrying about being too slow or too fast. You can wander crookedly up a mountain following only your whims, stopping to notice anything or nothing, and be free, absolutely free to think the most ridiculous, unconnected thoughts — and sometimes realize, at the top, that they were not unconnected after all, but have all come together in a single, inescapable point, a discovery.

I gather my gear together — knife, matches, water, raisins, and a book (Richard Yates's *Revolutionary Road*) — and start up into the mixed fir and spruce forest above camp. George left earlier, moving as slowly as if gut-shot, and after a while I come upon him, cross-legged at the edge of a cliff, staring into the churning waters of Rio Diablo. We talk for a bit to keep from being rude. He tells me a marten sidled up before I arrived and watched him for a few moments. It was the first marten he'd ever seen, and he described for me its long and heavy brown body, its little bear ears and bright eyes. It seemed to George that the marten accepted him as a fellow creature of the woods.

Martens are twelve to eighteen inches long, and they strike me as being a cross between little bears and big weasels. Martens can be predaceous or can feed on fir and spruce cones, like a squirrel. They're dependent on old-growth forests and have lovely fur; they are vulnerable to trapping. I remember the ancient skeletons Doug, Marty, and I found dangling from the trees two years ago.

"Maybe he'll come back," George says of his marten. I know that's my cue to move on, to vacate these healing grounds so George can turn his attention back to the rapids below and wait for the wild to return. I leave quietly, walking the game trail up the ridge through the trees. The forest is filled with the fresh smells of morning currents rising from the rapids below, and of sun-warmed fir needles and spruce sap and damp shadowy spaces.

It's the thirtieth of June and yet it feels like mid-autumn, with a north wind dropping in from over the summit. The sky's blue but the air is cold. I stop and rest by one of the many side canyons I have to cross, paralleling my way along Rio Diablo, moving up toward its snowmelt headwaters, and it seems this morning that even the cries and songs of the birds are different. Though the light is still bright summer sunlight, and the sun is as high in the sky as it is every June day at this time, the light has an austerity that threatens to burn a hole in the paper on which I'm writing, like light through a magnifying glass.

The infinite relationships of nature: if one hurls oneself over the cliff, away from the world of man and social graces and inanities — if one runs deep into the woods simply to wallow, root hog-like, in the natural world — might one's sensitivities to these variations be nurtured? Might a man or woman become more aware of strange things not readily noticed, much less explained?

What lies out there just above our heads in the spirit world, and just around the corner in the dark woods? What thrumming powers murmur beneath our feet? If there are men and women whose hearts and minds fill too full with the nearly infinite systems of artificial intelligence, then surely there are woods savages who similarly indulge themselves by rooting among the infinite systems of natural intelligence. Today I feel like one of those savages.

I have no idea why the light is so different today. Perhaps if I were to mention it to anyone else, he would think I was crazy, would insist that the world today is precisely as it was yesterday and will be tomorrow, the same as it has ever been, and that there is no variation, no mystery. And the scientist, as opposed to the witch doctor, might say that it is not any unusual change or rearrangement of molecules that gives startling clarity to the air and the light, but rather some chemical change inside me, in combination with the increase in elevation, that makes me perceive that the mountain sunlight is trying to burn a hole in the stark white paper and is trying to get me to slow down and notice still more strange things.

I drop down one of the canyons into the Diablo itself, a tortured nest of white cobble boulders, river rocks, braided channels with willow islands between its braids. Mica glitters in the sun, and chalcedony glows in the calm pools of eddies. I find an old hunting trail paralleling the river and follow that for a while.

Next to a brushy avalanche slope, I stop to eat an apple, and nestle down in a pocket of sand. I watch the clearing for a bear. Such a cold wind! Perhaps we'll get snowed on, this last day of June. I pull my paperback out and read for a while, barely able to stay awake. My eyelids grow heavy. I sleep lightly and for a short while, then the cold wakes me, but I feel immensely rested. I read some more, watch the tiny clearing, and then continue up the trail, daydreaming.

I think about all of the last killed bears — the Wiseman bear in 1979, the Lloyd Andersen bears in the 1950s, and the 1951 Al Lobato bear (the pitiful little head on display at Platoro Lodge). Sunk in the past, not really paying attention to the present save for the crispness of clean wind-scrubbed air and the way it tastes going in and out of my lungs, I'm surprised when I walk right into yet another abandoned hunters' camp. Tattered plastic sheeting still hangs askew here and there. Blackened aerosol cans of Cheez Whiz sit in the fire pit, which sits in the middle of the trail. Assorted Styro-ware. Rotten leather boots.

Where *are* these people? Are they back in civilization now, appearing to all observers to be as normal as pie, but inwardly ticking like time bombs and spreading their hot poisonous seed through the world like black-breathed plague?

I bolt from the trail and run down to the river's rushing edge, shuck off my boots and socks, roll up my pant legs, grab a walking stick, and wade across the knee-high rapids. On the other side I sit in the sun on the gravelly bank and dress again, then bolt up a nearly vertical avalanche chute. Enough monkeying around, I tell myself, enough gentlemanly sauntering. I will punish the memory of the cretins below with the wild hammerings of my heart, lungs, and legs. I'll blow a clot, burn out the impurities, die of a burst heart engine, or be saved. The blue sparks and lightning flashes scorch my vision once more, and still it's all a mystery to the doctors.

Straight up and up and up, like a small gorilla climbing the Empire State Building, only I am moving through old-growth Douglas fir, and the river becomes smaller and smaller below me. I'm drenched in sweat but still angry, so I keep climbing, sometimes hand over hand, to keep from slipping down the mountain.

It takes almost an hour before the memory is gone, slipping into the background like a shadow or a fading bruise — before the mountain has seeped into my back and legs, before its flora have calmed my skittering mind. I sit down on an elk switchback next to a giant fir in cool shade, and my peace returns.

I let the mountain — new to me, and me to it — transform itself, communicate itself into my legs, in the manner perhaps of a computer programmer speaking to silicon memory. The mountain is download-ing information and I sit and breathe it in.

The sun moves two notches to the west, ignites in brilliant light the gold fur and hair caught in the bark of the tree I'm leaning against, causes a yellow blaze in my vision; then the sun moves one more notch and the strange glow subsides. I reach up and pluck the hair from the fresh-rubbed sap. Most of it is elk hair, coarse and hollow, but some of it is downy and fine, like that of a grizzly. Would a bear rub its back on a tree that elk used as a rub to mark their territory? Would they share? It's a fine cool spot, with glimpses south through the fir boughs to Rio Diablo. Perhaps I am entering some spot in the forest where all the animals seem to congregate, for reasons not clear to humans, unless perhaps simply because the view is good.

I climb some more, a bit trembly-legged and a bit euphoric. The Douglas firs are still as big as redwoods, though the altitude must be about eleven thousand feet. Through the fir boughs, looking up, I catch glimpses of cold gray rock, the chalky slopes of the mountains' spine. I make my way over to one of the gorges that corkscrew down into Rio Diablo, a narrow, terrifying bare-rock chimney that leads all the way to the bottom. Across that gorge rises more chalk scree, snow skiffs, bare rock up to 12,500 feet. I am in the last and highest forest in this drainage. The Douglas firs are bigger even than they were before. They can be slow-growing trees, and it upends one's sense of order to find them here, at this elevation. This forest, these trees, rising so large through the centuries in the sheltered dark lee of the mountains, may be a thousand years old.

Sunlight illuminates various works of perfection — the rotting corpse of a fallen fir, a gray fieldstone spangled with lichens, a bed of ferns. I look upslope, and freeze when I spot a four-legged silhouette above me — something larger than I am, and as startled by me as I am by it. Somehow I sense, by the way it is standing stock still, that this animal has never seen a human before.

A mule-deer doe. Larger than a pony, she's on a little flat spot in the forest, the last level ground before the slope flexes again and rides on up to bare rock. I step closer in the forest dimness and she canters left, toward me, then away, and I realize that she is confused, acting in the way I've seen deer and other prey act when they're being stalked or chased by predators.

Another doe appears to my right, and she too seems indecisive, panicky. The two deer whirl and spin, toward me and then away, like synchronized swimmers, flowing in unison in their distress. Why don't they just bound away, in high pogo hops? Why all this *thinking?* I see plenty of escape avenues available to them. Because the deer will not run, I move closer, in a human way, to see how close I can get.

I have hunted deer all my life. I live in a valley that, owing to excessive clear-cutting, is overpopulated with deer, and underpopulated with predators. I see perhaps ten thousand deer a year, and perhaps fifty bears. I grew up in the Texas hill country, another place overrun with deer. Their habits and movements have become as natural to me as air or water or clouds. There's something going on in the forest now, something in the air, a ringing strangeness. I move toward the deer like a sleepwalker.

A great wind-weathered fallen fir tree lies on its side halfway between me and the skittery does, which are now only thirty yards away. When I am ten yards from that fallen tree — which I am all but ignoring, focusing on the deer — a creature leaps up from behind it, seemingly right in my face, a brown creature with great hunched shoulders. It's a bear with a big head, and for the smallest fraction of time our eyes meet. The bear's round brown eyes are wild in alarm, and mine the same or larger, I'm sure. The bear's a rich chocolate color, like a moose and nearly as big, an animal of such immense size that indeed my first thought, the one right before fear, is: *That bear's as big as a moose!*

An awe, a reverence, nearly takes seed, the idea that here on this highest reach of mountain a bear can live to reach such an immense size. But the reverence flees immediately, obscured by my desire for escape and safety as I look for a tree to climb, my heart in my throat. That glimpse of the rolling humped back and the wild, wild eyes is all I get before the bear's flight takes it down to a wooded ravine and away to the left, into a sun-filled avalanche chute.

The tree I have in mind to climb, the only one small enough for me to have even a remote chance of climbing, lies three steps in front of me — toward the bear! I take those three steps in a hurry, despite all knowledge and warnings against "making like prey." I can't help myself: my body overrides my mind. Meanwhile, the deer have exploded to the right, over to the gorge's edge and down the rim of it, the plunging line where land seems to disappear.

By the time I reach my tree — aware, in the course of three steps, of the ludicrous paradox of running *toward* a bear — I realize that if the bear were after me, I would already be cornered: the bear would have charged and engaged me by now. I lean against the tree and listen to the fade-away sounds of occasional branches breaking, then silence — grasses and ferns underfoot, I suppose — then sounds of sliding shale, and then nothing except for my heart and the electric pulse of my blood.

Curiously, giddily, the George Bernard Shaw epigraph to George Plimpton's *Shadow Box* comes to mind — "He was not raised in the forest to be frightened by an owl" — and after a minute or two I know my duty. Although I have been thinking *deer,* I am up here to look for *bear,* and I walk with a perverse exhilaration toward that fern-filled ravine, not unlike the prisoner at sea who finds himself walking the plank on a fine spring day and enjoying the weather.

The sounds the bear made "escaping" came from below. When I get to the edge of the woods and look into and across the tiny sun-spangled slope of green — bear heaven, with a spring trickling through it — I notice that there is a mule-deer buck standing in the ferns slightly upslope of me, looking dazed, as if he's seen more action in the forest this afternoon than in all of the previous year's afternoons combined.

The buck, a light tan color, watches me with huge wet dark brown eyes, eyes of fear, not unlike the bear's, but even larger — the eyes of prey, not predator. And most striking, most beautiful is the immense rocking-chair rack of velvet antlers the buck is wearing, a crown so large it seems the buck can barely carry them.

The mule deer spies me, makes his decision to leave, and bounds past me down the slope in that jumping prance that mule deer always use, all four hoofs hitting the ground at the same time, *boing boing boing,* sending the concussive and distinctive sound through the woods, a sound I did not hear when I jumped the bear from behind the log.

The bear had been stalking those deer, trying to corral them against the rocks. (Earlier in the climb, a couple of hundred feet lower, I had smelled but been unable to find a dead deer.) The deer had been nervous to begin with, and then doubly confused by my arrival.

I had walked into the middle of "something," rather than the usual, seemingly slow "nothing" of nature. Many times during a predator's final charge on its prey, the commitment has been so great on the part of the predator that even the surprising arrival of humans on the scene won't deter the completion of the act. I'm convinced that was how I was able to get so unnaturally close to the bear.

Which I believe was — is — a grizzly.

The hawk that stoops to snatch a ground squirrel on the trail, directly in front of a startled hiker; the coyotes that chase a hiker's dog right between the legs of the dog's owner; the snake that cannot be turned away from the fish on the fisherman's stringer; the buck deer that, in rut, runs straight through the hunter's camp in pursuit of a doe — there are in nature certain cycles and pathways so integral and highly evolved at their outer reaches that, just prior to their moment of completion, no force is capable of stopping them. I have stumbled by accident upon such a cycle. How I was drawn here — the sequence of events that conspired to put me in this special forest and this exact moment — is another story entirely. (Down on Rio Diablo, should I have read another chapter or not? Should I have slept a few minutes longer? Chatted with George a bit longer? Fled the deserted camp in such a direct, vertical direction, or one more oblique?)

Still shaken, I stand at the edge of the chute's clearing and am filled with an inexplicable sense of loss. I want *more.* The emptiness of the avalanche chute seems too huge, and the briefness of the encounter too severe — a glimpse, the meeting of eyes, the bear's head and upper body rising and then disappearing below the great tree in a rolling sort of surge, then nothing more.

I saw a huge bear. Not a tan deer with antlers glowing in velvet. I saw a chocolate-brown, round-headed, humped-back bear. Like a predator, moving toward those does, I had been thinking *deer,* but what leapt before me, almost right in my face, was a bear.

The anticlimax of it — the absence, the continued lack of proof — is so strange and disheartening as to be almost crushing. What else is there to do at this point but to protect myself, to create doubt within myself, perhaps to keep the notion of the search going in my mind, rather than a two-second glimpse that ends it, ends it up near twelve thousand feet?

I want to hold something. I want to have my picture, anyone's picture, taken next to the bear. I want the bear to return from the woods, loll in the meadow, roll over on its back and say, Okay, the jig's up, here I am.

Ashamed of the lack of allegiance I have to my own eyes, I walk over to the great fir tree the bear had been hiding behind, crouched, waiting to ambush the deer herd before my blunt and blind arrival messed everything up.

I can find no impression in the duff. No hair, no scat. I want to wail.

Perhaps I am growing sick with the steepness of the mountain and the gain in altitude, but I cannot shake the thought from my mind, the shameful thought, that *it does not matter:* that what happened today was between one man and one animal and of no more significance than a sparrow taking flight at the approach of a man's footfall, or an aspen leaf landing in a stream and being swirled along down current.

There's nothing to do but climb. Farther into these strange woods I go, spooking a large boreal owl out of a big fir. Up a wooded cleft of ravine, beautiful shadows and sunlight, tall trees, doubting the secret in my heart. At the edge of the rocks I find coyote scat, bleached white and containing deer hair. Every creature in the forest seems to prefer this lovely stand, this elegant little forest. Now I start up into the rocks, for there is nothing else to do now but climb, as if God will invite me in for a cup of tea and answer my question: Yes, Rick, you saw a grizzly, a big dark brown male, which saved you — a sow with cubs would have eaten you and left you as spoor all over the mountain.

The cliffs are steep and loose. In one rock cove not far above the tree line, I find a slender brown canid-like scat, so greasy and foul smelling that I suspect wolverine, but I dare not put it in my pack. I see no hairs in it. Perhaps the wolverine, or whatever creature left it, ate only the guts of the kill — definitely a stinky meat-eater's feces. Demoralized still by what I'm feeling as my loss rather than my gain, I climb on, up into the cold winds.

The magic forest grows smaller below me. Dizzying near-vertical slopes of gray rock and chalky rubble — remnants of old oceans, diatoms' calcium skeletons — tower above me in their new home in the sky. It's a long way down. I keep climbing without knowing why, until it occurs to me that what I am doing is dangerous, that if I slip or fall there's nothing to grab on to, and I'll go all the way to the bottom. Even as I'm registering this oxygen-thin thought, the chalk gravel shifts beneath my boots and I lean in closer against the mountain's face.

I have a daughter, a wife, a family! Why am I continuing to climb this ridiculous sliding sheer wall, pebble-sized talus, just barely clinging to the mountain's slick face? Why? To get away from the truth?

I saw a bear. I believe it was a grizzly.

It was a magical experience, but like a Puritan, despite all my fond celebrations of the importance of mystery, I realize that I am afraid of it, afraid of mystery and magic. I do not think the bear — Old Grandfather, Illustrious Master, Honey Paw — revealed himself to me. I think the mountain revealed the bear to me.

I find a small cave pocked into the side of the cliff, beneath an overhang, and lean inside it, sit down and open my pack for more food: a peanut butter and jelly sandwich and more raisins. I listen to the wind, and, as if there has been a slippage in time where my heart has slowed to a pace at which I can better feel and understand things, it seems that I can feel these young mountains growing slowly beneath me. I can feel them swelling above the tectonic plates of their birth, and yet simultaneously being stripped down from above by the quick, furious passions of weather — by lightning, frost and thaw, wind and snow, and the seasons: the mountains both growing and sinking at the same time — *pulsing* — like any one of our lives. I feel the altitude working on my body, changing the pounding of my heart and my mind, bringing me almost perfect stillness.

What I was given a glimpse of — and so close, close enough almost to touch — only now begins to settle in, to expand like cracks from a tree's root, expanding within a boulder. I am so rattled by the sanctity and strangeness of the encounter that I continue to try and construct in my mind ways for it *not* to have been a bear — for it to have been a figment of my imagination, the hallucinatory tremblings of an aneurysm, perhaps. The Blue Spark Special. Are the things we see and feel real? Aren't they as real as rock? We *are* still alive, still sensate, aren't we?

It *was* a bear, a lunging, fleeing, giant bear.

I came in prepared never to see the bear, never to find sign. I came in prepared to be protected by failure, to admit loss. In no way that I can think of did I do anything to warrant the mountain revealing its true power to me.

It occurs to me that a two-second glimpse of the Old Man, in his magic forest at the top of the mountain, is just about right, just about correlative for a sighting of the most keystone of species. Everything below the keystone species is a form of lichen. Doubtless, to the mountain, the bear is a sort of glorified lichen, and to God or Wakan Taka or Allah above, the mountains are themselves lichen — slow-growing things that flare up brilliantly, as if after a rain, but that settle back into long periods of dormancy, disappearing beneath oceans for millennia.

The utter holiness of being alive and part of such a system, the holiness of being allowed to be a lichen within the system — I'm not normally a cheery person, but here on the chalk-rubble slope, tucked into my little lunch cave, I find myself grinning, then laughing at how tenuously alive I am. To hell with electricity, with sizzling nerve endings and mispronounced words. I want to learn a new language anyway, the language of breathing forests, the language of further mystery.

Humbled and still a little frightened of this manifestation of power, but happy, I pack my things together and start down the treacherous slope, deciding not to attempt the last two hundred feet above me, which rise even straighter and more dangerously. I am a woods walker, not a mountain climber, and the green forest far below beckons.

Rocks trickle from beneath my boots, bounce in the long journey down to their new resting spots, raising white plumes of chalk smoke, and I'm embarrassed, one man making so much noise and disturbance on the mountain.

I make my way gingerly down the cliffs and finally reach the tree line. I turn and look back up at the shining white wall above me, a little frightened and a little pleased at how dangerous it was. I walk over and inspect the violent vertical beginnings of that corkscrew gorge, the slot in the earth that drops almost two thousand feet down into Rio Diablo's fast wet belly.

An enthusiastic disciple of Doug's "never go back the way you came in" philosophy, I aim in a direct line for the river below, not worrying about which particular game trail, if any, I'll intersect — just trotting and then galloping down the steep slope, yielding to gravity. I'm popping branches, landing heavily when I vault logs, careening off smaller trees. It's more of a tumbling down than a true flow, but still it feels graceful, as if something in the mountain and something in me are connecting and binding, even if only for the duration of my flight. The speed, the flow, stops abruptly when I spy below me the fast-approaching lip of a sheer jump-off. This is not the way I came up, I think, clutching at ripping-by saplings and veering ankle-bending hard to my right, cutting up duff and forest sod but still not stopping, still being carried — *thrown* — down toward the cliff's edge, as if a great hand is pushing on me from behind.

But right before the cliff's edge, the mountain levels out a bit, becomes nearly flat for a width of perhaps twenty yards. My momentum carries me all the way to the flat spot, where my feet and ankles are able to resume fuller cooperation with the traditional laws of physics; my boots find hard purchase. I whiz along the side of the cliff in a tight arc, still running, paralleling it now, trying to slow down. I clutch a medium-sized fir at the edge and wrap both arms around it.

I peer over the edge, let go of the tree, turn and walk carefully away, back across a marshy, boggy flat spot. I find a spring seeping out of the mountain at the top of a fault and sit down, trembling, and gaze out at the space beyond.

When the trembling is gone I walk back to the edge, hold on to the same tree, and look down at the impossible jungle below. I feel the breeze lifting up that cliff, still driven by the heat of the day. Farther to the right, I can see another avalanche chute. I walk along the meadow seep at the top of the cliff, sinking heel-deep in muck, and search for tracks. I see nothing but elk sign.

I cross a tiny creek, drop down through timber to the next level of cliff, and find myself following the same game trail on which I'd come. I reach a switchback I remember pausing at on the way up. A distinctive elk turd rests atop a flat rock, the scat made memorable because of the remarkably undigested Heracleum leaf lodged in its middle, curling out of the scat as if still growing. It was at this point, I remember, that I had left the trail and struck straight for the top, because of the relatively gentle slope of the switchback.

It is getting late in the day. I'll just follow my old path back down a ways, I tell myself.

I haven't gone ten steps when I find fresh bear scat on the side and just a little above the trail. An immense quantity of it, in fact, as thick around as my arm, and coiled, fresh, and full of grass bits and nuts and seeds. It is a scat, as Dave Petersen will later put it, indelicately but accurately, "about the size of a dinner plate."

Thoughtlessly, I stuff the scat in a plastic bread bag that I carried my lunch in and load its significant heft into my pack. It will not occur to me until later that this is the scat from the bear that I "bumped" several hours ago. Because the scat was right by the trail, I would definitely have seen it on the way up if it had been there then.

There is another spring above an avalanche chute, and I walk over there to look for tracks. I'm a little red-eyed and sloppy-footed from the day's rigors, so I don't see the second pile of scat until I almost step in it. It, too, would fill a dinner plate. I wonder, Do I have to carry it all down? How about just a sample? I wrap up a good bit of it in a plastic bag I keep in my pack for rain wear and tie it off.

I examine the spring for tracks, but can find none. The weather probably isn't warm enough for the bear to roll in the mud, and I remember Peacock telling me that bears are often not much different from people: that some will tromp right through the middle of a puddle, while others will go out of their way to stay out of it.

I cross the spring, stepping from fallen log to fallen log, to avoid the mud. Less than a minute later, bushwhacking now, I come across a third scat, which is every bit as enormous as the other two. All three are damp-fresh, and all were left within a thirty-yard radius. I pull out my rain bag, which sags with the weight of the second scat, and fold it back on itself — the knot is tied in the middle — and in the upper compartment formed by the knot, I load the third scat and tie it off.

I can barely fit it all into my day pack. I have no more plastic bags. If I find any more scat, I'll have to carry it down in my hands. I am at 11,200 feet, still about three miles from camp, and I do not want to find any more. I've never seen scat half this big, not in Alaska or Montana. If this isn't grizzly scat, it must have come from an elephant.

I drop straight down the tangled edge of the avalanche chute with my shit pack heavy on my back. Sticks poke me in the ankles, fir branches pop me in the face. Down toward the river's roar with that incredible feeling of extravagance, giving up all of the day's hard-earned altitude, giving it up so freely and completely. Spend it all, writes Annie Dillard, spend it all. Every last drop of it — the juice, the passion, and the elevation of the mountain. Having taken every foot of it into your body on the way up, the only way to hold the mountain in you now is to release it, to give back to the mountain every inch that you hoarded on the way up, and not to be shy about it either.

Twilight is filling the narrow canyon. At the river, I find myself in a wider, faster, more violent section of rapids, and must strip and toss my clothes and shit bag across the rapids and then ford the river with only a walking stick.

The river rocks punch my feet. The delicious shudder of the river rushes around my legs as I make my slow, old man's way across from the side of the bear to the side of man, water splashing my butt and balls, and then I am on the gravel and sand on the other side. I sit on a boulder for as long as I can stand it, drying out, shivering in the strong canyon wind. I put on my clothes, reveling in the dryness of dry wool socks over my feet, and start up into the woods, crossing ridges, my back to the bear now, headed toward my friends.

I reach camp in heavy twilight, with a ghostly mist rising from the iris meadow. I set out the scat on the black tarp and am oddly embarrassed by how my weighty loads dwarf the other scats there.

It's been a strange day; to describe it, I think of the old-fashioned word "authentic." Tolisano and the others are excited by the scat — a little surprised, as I am, that it can be that large. Now I wish I'd stayed up around the spring to look for even more.

It has been an authentic day. Dave found three small scats near the iris meadow. The student Jim Sharman found a single wisp of blond animal hair on a bush beside a trail — how he noticed it amazes us all. It looks like grizzly hair, a couple of inches long, silver-tipped, and blond all the way through the roots. And more good news: George looks rested, more like his old self, capable once again of being engaged with the real world. He comes over, inspects the scat, and seems jolted by it. He smiles slowly.

We all stand around and stare at this mystery that has come down from the mountain and into our camp.

Jim Tolisano's talk that night is on the subject of land ethics.

"Most people," Jim says, "no longer have a feeling for the place they live. How do you deal with people, then?" he asks us. How do you deal with the people who drive to work, work in buildings without windows, then drive home, eat, and go to sleep?

Much of success in nature has to do with finding niches and with the competition — muscling in and imposing the sheer will of one's wishes within a system. Jim counsels us to take a more humanistic approach, to try instead to find common interests in the community and then nurture them, to develop and strengthen a bond in any area of overlapping interests. "Number one, we define our purpose. Two, we find out the local communities' needs. Three, we focus on where the overlap is between our purpose and their needs." In that manner, Round River hopes that the communities will get involved, will help give positive and directive force to the San Juans' future and protection.

"We can talk to people in the towns," Steve says, "in the bars and Laundromats."

Several of us — the older ones, I notice — study the fire for a while. We've been in towns; up here is where we want to be. Circulating among strangers and spending time with concrete under our feet when there are perhaps only a few years left does not sound very attractive to us. But we know Steve's right.

The younger Round River folks nod with hope and enthusiasm. Part of the curriculum calls for them to write an article about the summer's experience and submit it for publication in their hometown newspaper. They don't yet have scars across their hides from brawls over places they love. They're not naive, but watching their eager eyes around the campfire, watching them doodle in the dirt with crooked twigs, sitting cross-legged before the fire and in love with these glaciers and rivers and bears, makes me feel both tired and encouraged. They think it's going to be so easy, I tell myself. And with Tolisano at the helm, with Sizemore doing the thinking, and with Peacock behind them, pushing the boulder, his raw explosive fuel of passion for this dream's journey, perhaps it will be easy.

The night deepens. George Fischer has begun speaking again, telling of how he would sometimes go camping on a gunnery range in the desert west of Salt Lake City, about the strange lights he'd see. He says the Air Force tested all kinds of aircraft out there late at night. One Stealth bomber went right over his camp once, flooded him with blinding light. It paused, then moved on — the Air Force somehow more terrifying than any Martian.

"They could have shot you," I say.

George shrugs. "It was winter. They didn't figure anyone would be out there."

I marvel at the coincidences that place people's lives at intersections with certain events. I am thinking of course about the bear up on the mountain, perhaps three or four miles upstream, and when Dave brings up the subject of the scat's immensity, I see that I have to testify.

"I didn't want to say anything," I tell them, "because I thought it would make me look goofy, doubting what I saw." I go through the whole story again.

Dave, who hunts more than I do, insists it was not a mule deer. "They always hop when they're frightened," he says, when I make clear that this huge brown creature lunged and rolled, then *ran*. "And you would have seen the antlers," he says. "No way would you miss the antlers."

What I saw was a bear. But I have to confess to my fellow soldiers at least a fraction of the confusion I feel: how I've been so geared to bringing in the tactile — the hair-filled scat, the claw marks on a tree, the footprint in the mud — that to come in with the surest and yet most ethereal thing, a memory, well, it confuses my gears, upsets the natural order in which I've been operating. I can't help but feel that I've let everyone down by not bringing the bear back into camp, or camera footage at the least.

I shouldn't have told the truth right after George's story — should have kept it in my heart — for now someone is talking about a UFO.

If it were not for the scat, I might believe that I had been in the high-mountain presence of a deer, and the deer watching me then with only the brown eyes of a bear. But the deer was above, and what I saw was chocolate-colored, not mule-deer tan. What I saw lurched, rolled, and ran like a bear, and had been making those deer nervous. And what I saw in those eyes, that brief wild meeting of eyes, was fright, looking so much like the frightened eyes of man that I was granted the beginning of a new perception. I will never forget those two seconds, those short turns of the gear of the day. What I saw was not for scientists, or even for Round River, but for me — and even that was by accident.

The mysticism of the event is what leads me to believe it was a grizzly. The strangeness, the power of all that surrounded it.

*

Having run low on food a couple of days ago — young people eat so much, and young people climbing mountains, the most — the Round River students have pooled the last of their pitiful stock, half a bag of wheat flour, and have added water to it, stirred it up, and poured it into a dutch oven in an attempt to make a sort of wheat pie. It's been baking in the flames for a terribly long time — there's about two gallons of it — and all night stomachs have been grumbling.

Because we cannot eat, we talk. Stories of danger, rather than those of unreliable sightings, become the theme. Scott Carrier and Jim Sharman are giggling about how they nearly slid down and over a glacier two days ago, up on the Continental Divide. "It was *close*," Scott says, laughing, and Tolisano shuts his eyes, tries to put the image out of his mind. This reminds Beth of how Scott was crossing a log over some rapids, earlier in the trip, and the log broke and he fell in, pack and all, and would have been swept away had he not grabbed on to a branch at the last second. At the time, Beth laughed; couldn't help herself, she says, because it looked so funny, even though she knew he was in danger.

This brings up tales of lightning. The whole crew had run down the spine of the Divide in a hailstorm, pellets bouncing off them, wind whipping their faces, the smell of ozone in the air, and jagged lightning strikes all around them. They did this because, as one of them put it, "We had never experienced that before." Tolisano closed his eyes again and hid his face in his hands.

The wheat pie is ready. At least, some consensus over this fact arranges itself out of thin air, it seems, because to me the pie should have been ready hours ago. At any rate, a vote is taken and — true mystics! — everyone decides that yes, now is when the dish must be ready.

With maximum barbarity, the heavy dutch oven is maneuvered out of the flames with the help of sticks and logs. Dirty socks are employed as pot holders. The lid is lifted, like a vault with treasure inside. Black smoke pours out of the skillet as if the lid of hell has been opened, and the night fills quickly with a scent that immediately evokes the gag response.

I can't help it. "Shit, Jim," I say to Tolisano, "I know you said they were dangerous, that they went out on glaciers and fell in rivers and stuff, but you didn't tell me they cooked, too."

The black soufflé, light as cardboard, is pried out with a pocketknife.

The blackened disk is passed around the campfire, and we gnaw on it like rodents. When it is my turn at it, I am unable to shake from my mind the idea that we might as well be eating charcoal briquets.

But they're young, unsinkable. "Not bad," Chad says, chewing, his words barely audible over the gnashing of his teeth. "Not *too* bad." And wouldn't that be a hell of a thing to learn on this trip, that indeed nothing truly is *too* bad.

One of the major failings of our own species — also a keystone indicator — is that we learn stuff but then forget it. As a tender disciple to mystery, I prefer sometimes going back after that which we've forgotten, as opposed to charging off full tilt after something new. "Never leave a man behind," soldiers in battle tell one another, and I feel the same way about the San Juans and all wilderness. Let's take care of and protect our wild past and remember everything within it, better to strengthen us for our race off into brave new worlds.

Power, earth, senses, lust; grease, fat, food. After the sad pie is gone, rocky in the empty bottoms of our stomachs, we indulge in the almost obscene enumeration of the foods we crave once we get out, describing in detail what we are going to do to them, the slow pleasant manner in which they will be ravaged.

"Ice cream . . . dripping . . . watermelon . . . sucking . . ."

"Big, fat, juicy, dripping raw red *cheeseburgers,*" says the lean and spartan Tolisano, urging on the frenzy. Chad, poor Chad, rolls his eyes back and appears to be having a seizure.

*

I love the way campfires break up — the slow release of that bond, the flames becoming coals, becoming almost nothing, relinquishing us back to the night, back to the stars, so that when you look up at them, you seem to be among them, like a giant, or like a cinder that has floated so far and high above the earth.

After midnight, the students drift off. Not until they are gone do I realize what the reason for it is: the older guys are talking gloom and doom.

Only one of the students, Dan, remains to hear George, Scott, Jim, Dave, and I bitch and moan. Dan is from San Francisco, and with his carrot hair, and his restless kind of power, he reminds me of Peacock. Dan's familiar with the environmental legislation before Congress, and he knows who's who on House subcommittees. He knows the key members of Congress who are deciding Montana's future, and he's also up on the Bureau of Land Management's wilderness activities — or lack thereof — in Nevada and Utah. He's got the energy we need, "we" being the future of Western civilization. Everyone's been talking about what a terror Dan is on the trail, how he never gets tired. We can see it here under the stars, after the fire is gone.

The night is cold, and we see Dan sort of stepping up and then back, up and back, walking in place and sipping his cold coffee, talking about all these things with us. It is taking the edge off our despair, and it's the reason why, when four or five of us get together, we feel like we can change things. I'll never know why, but it happens, and we feel it once again that night, there under the stars.

*

The last morning. I sense that the meadow will be glad to see us go, although while we were there, I did not feel that it found us unwelcome. We spread our campfire ashes in the woods and carry our fire-ring rocks to the edge of the cliff and toss them over the edge. I'm amazed at how sharp and simple the pleasure is, how primitive and secure, watching the rock bounce from ledge to ledge, all the way down to the river, while we stay safe up above, at a distance from harm's way.

For breakfast, the students split a banana eleven ways. I have slipped an extra can of Vienna sausages to Chad on the sly — the first meat he's had in three days, this bench-pressing fur-trapping wrestler from Ohio. I remember Peacock's and Terry Tempest Williams's words one spring: "The best we can do is to cook for one another." I'm not a good cook, but I can haul groceries. After our skimpy meal, we gather in a loose pack train, shoulder our packs, and prepare to descend the ratty yellow rope. Dan will go last, bringing the rope with him.

"What'll we have for supper tonight?" Tolisano asks, rubbing his hands together and grinning.

"What you said last night sounds good," Chad mumbles shyly.

"*Cheeseburgers?*" Tolisano says, and Chad smiles.

Steve's boots have blown out over the course of the trip. He's had to hold them together with duct tape, so he looks like a hot-shot football player from the seventies. Until the duct tape was found, there had been talk of trying to use some of last night's uncooked wheat-pie dough as glue.

At the first stream crossing we take off our boots and socks, roll our pants up, and unhook our packs' hip belts. We cross gingerly over the pointed rocks — such a young stream. The little canyon is filled with the sound of our bellyaching, and then our slow, blissful silences as we reach the other side and sit on various boulders, massaging our numb feet. "It's easy to confuse pain with warmth," I call out cryptically to those who are still crossing the fast stream, hunched over and whimpering.

The game trails are deep, as if worn down by heavy horses' hoofs. Actually, large herds of elk have cut into the slopes, moving up and down the same trail day and night. It's easy walking, but even on the steady drop of slope, freighted with my pack, I feel like a slow Neanderthal. I look at the woods below me, logging into the mind's grid the changing flow of new information that tells me where deer and elk might be, and mushrooms, springs, bears, everything. I process the view with my dim and dusty electrical circuits. I try to think like a predator, and my mind can do it, but not my body — too slow!

A theory by a British ecologist, Allan Savory, suggesting that predators prevent erosion, has been in the news a lot lately. More controversial is his theory that the land can sometimes hold more ungulates than we have currently, if there were more predators.

In Africa, Savory says, the great ungulates move in large herds as a means of protection against predators. In this country, before the bison were gone, we had a similar system: herds numbering in the tens of thousands. As in Africa, our great herds came into new grazing grounds, spent two or three days there, and then, aided by the attentions of predators — lions in Africa and wolves in this country — moved on. In this manner the grass was grazed intensely, but over a very short period of time. This served to invigorate the grass, encouraging healthy new growth the way pruning can do.

But, says Savory, with the predators gone, the prey became sedentary and developed daily habits uninfluenced by ambush. Herd animals used the same trails again and again, staying in one place too long and grazing grasses down to the roots, causing erosion and loss of rangeland. Trails were cut deep into the earth.

Which is harder on the land, he asks, sending a hundred mules down a hill to get water on one day of the year, or sending one mule up and down the same trail once a day for a hundred days? With predators — with a responsibility to adapt and avoid predation — the prey would disperse.

It sounds right. It's plains ecology, not woodland ecology, but it sounds right.

Colorado's elk populations continue to reach new all-time highs each year. I look at the elk trail we're following — a wide superhighway through the wilderness — and it's easy to see the localized erosion already. It's easy to imagine the manner in which that localized erosion will spread to general erosion: loosening roots causing the loss of topsoil, loosening more roots; then a slump, a mudslide, and all lost; another slump, conchoidal slumping of the mountain, revealing each time bare soil, which natural forces tear at eagerly, wind and rain and frost, the newly exposed surfaces having little or no resistance . . . All because there was not a wolf.

One wolf pack can save a mountain, if Savory is correct. One mountain lion, one black bear, one grizzly. One wolverine. What the mountain really fears, says Aldo Leopold, is a lack of wolves.

Tolisano's right: your mind can explode trying to understand all the data, but when you use the elementary building-block formulas of keystone species and indicator species, it becomes very simple. One wolf equals one mountain. One mountain can teach, or save, one man. One creek can be home to one family of otters.

Like children, we begin to stack the blocks and begin to learn, and relearn, our place in the world. Not above it or below it, but in it.

∗

Tough hiking, this pace that alternates between roaring along elk trails, stopping and climbing elk trails, then stopping and climbing down and up side drainages. Sometimes we stop and rest. On one side hill, perched in the sun, I feel freighted with lead. I know that once I get going I'll be fine, but all I want to do is *sit* here with my leaden butt.

Everyone's rising, though. It seems that, far ahead, Tolisano has started moving again. What I need is a hand up, or the offer of one, but the students haven't been backpacking much, and they just tramp by. I rise on old knees and, once up, it's okay. More than okay — it's glorious.

∗

Despite our pack train moving through the woods — all these *journalists,* one for every student almost! — we see some wildlife: a mule-deer fawn, fuzzy as a bull calf, spooks from the bushes and runs dapple-spotted right into our midst, zigging and zagging toward us as if believing we're a herd. This causes much excitement on our part, especially for George, who, though rested, still seems a little whacked out from his computer screen. He's wild-eyed after the fawn has hopped away. "You know how when something's coming straight at you and you can't see it?" he says. I don't really know what he's talking about, but I nod because he looks so shell-shocked. "You know how sometimes when you get the wrong thing in your mind, how even when you know it's the wrong thing, you can't get it out?"

Again I nod, and I think, George, you're not ready to come out . . .

George begins to giggle. "I thought it was a warthog coming at us," he says.

Big George! Neither warthogs nor mental breakdowns faze him. He has the ability to marvel at everything and to laugh at himself.

Down into the low country, then. The last of the side drainages have been hopped, and the canyon splays before us. We find one last bear scat, a small one, in the center of the trail. Dutifully we collect it, though we seem like different creatures now, more human. Collecting this scat down low, on the way out, feels only a little different from running to town on a mundane errand. The glory of high places is gone, and in its absence we must now work to reenter the glory of low places.

We stop just above a gravel wash. Very soon we'll be out; we've just crossed an old fence and an old logging road on private property. Beth sits down casually on the edge of a thin stump, her legs out wide, balanced there like a T-square. Her face is pale, and she's sweating prodigiously. She looks like I felt a ways back when I'd been longing for a hand up on the trail. In a moment of bonhomie, team spirit, that kind of thing, I offer her a helping hand as our pack train begins to move once more.

*

We've reached the Rio Diablo. Now we must only cross it and we're back to the dirt road that leads out. The Round River crew heads upriver, looking for a log to cross; Scott, George, and I take our boots off and ford right where we are. The river's wide and shallow this far downstream. There used to be an old bridge here, but floods have washed it out — only concrete piles and twisted cables remain. Once we're across the river, we sit on old timbers and let our feet dry out, scrunch them in the white sand, and absorb all we can of the mild sun.

It's cold and windy. George glances upstream to see if the rest of the crew has moved out of sight, and they have. He strips, wades back into the river, and lies down in the shallow rushing waters. Scott does the same. I sit on the log and listen to them gasp and sputter as if they are being skinned alive. After a few seconds, they scramble out. They walk around in circles in the sand, flapping their arms to dry off as the wind howls. It's a startling sight to me, and doubtless even more so to the old woman who's hiding, watching us from behind the bushes.

We're trespassing, it seems. We've crossed back into that system. After George and Scott have dressed, the woman emerges from where she's been hiding and explains things to us. She owns the land and has a cabin back in the woods. With utmost courtesy, but no apologies, we move along.

While it's true that naked strangers shouldn't dance in other people's yards, and that perhaps we're being overly sensitive when asked to move on, it occurs to me nonetheless that her proximity to the river and to federal wildlands allows her to enjoy a wealth of intangible benefits that we as citizens provide her with our taxes. Would it be any different if a bear had come out of the river and revealed itself? The idea of borders without confrontation seems incapable of being grasped by the mind of the terrible flat-faced bear.

Barry Lopez is right: generosity is an act of courage. How brave are we in what is left of the twentieth century? As a nation, are we becoming braver or more cowardly?

We walk along dusty dirt roads through new developments of summer homes with buck-and-rail fences protecting the gone-away owners' yards. Back at our parked truck, George talks about the night around a campfire last year when Peacock, joyously drunk and sated on chanterelles, fell on him.

"It was like he was made of iron," George says. "It felt like about four or five hundred pounds of iron. He didn't feel like a real person. I couldn't budge him off of me, either."

For the thousandth time, I ask myself what I'm doing in Colorado when there are so many other battles raging. I can't answer, but know only that it has something to do with the iron nature of bears, and the family of the West. It started with Doug, of course. I think that among our habitat of humans he is the ultimate indicator species, and that for all of us, East and West, whether we realize it or not, his battles are our battles. It has to do with all the old words: loyalty, friendship, heart, passion, survival. Iron.

We start the familiar drive back across the Southwest's high plains. The blue mountains of the north, and home, and cold winds and snow pull me back to the place where I feel strong, a place that is endangered in a similar way. I cannot shake the feeling that in this fight I've come so far out of the woods, drawn out farther than I ever intended. I want to help everything; whether I want to help too much, I don't know. It is like the feeling of having a very large, very tight family. And who knows? Perhaps someday I will not have that feeling of vulnerability, traveling from the Montana-Canada border down to southern Colorado; perhaps habitat corridors will protect safe passages for the genetic flow — the future and hope and passion — of wild things from north to south, south to north.

"They are a race that has been here forever," Doug has said of the grizzlies. "And to eliminate that race would be an unspeakable tragedy. This whole thing — Citizens' Search and Round River — is about human beings finding their own place in the natural world.

"It speaks directly to arrogance. If human beings are going to be more than a blip on the goddamned evolutionary screen, we're going to have to learn to accommodate the other residents of the earth."

*

We eat the double greaseburger at the Elkhorn in Pagosa Springs, and then stop just a few blocks away, at the hot springs. Scott and George know how to travel slow. At this rate, we'll get back to Salt Lake around daylight tomorrow — which is when Scott's plane will take him to an assignment in Washington.

An hour later, in Durango, we stop to eat again, because there is a good Mexican restaurant there. We're still full from Pagosa, but we sit down and have a bowl of posole, just to cleanse the palate, and order the rest of the meal to go, to ride with us in the back seat until we get hungry again.

We drive on. Food and wilderness, passion and lust, joy for living — it's all part of the West, and until the West becomes like any other place, or like one place, it will remain that way — a place where, as in the system of nature, times of spartan hardship ebb and flow, right next to times of joyous, even gluttonous abundance: a cycle.

We cruise across the high desert at forty-five, fifty miles an hour, as if it is only the middle of the century and not the end. North into Utah. Night is falling, shrinking the outside world, but also enlarging it by increasing mystery. Scott and George talk about computers and virtual reality. As I understand it, VR trades real life for the illusion of real life. For some people, this trade is actually stimulating, an improvement, an escape.

In the *Washington Post* Mark Potts writes, "Wearing a helmet that provides a television view and stereo sound, and guiding motion with a joystick or other device, a user can move electronically through a simulation of real life — or through something abstract that only a computer can create. . . .

"Turn your head and the scene changes accordingly, just as it would in real life. Put your hand into a special electronic glove, and you can 'pick up' items you see . . ."

I'm thinking about Tolisano explaining the futile attempt to understand the myriad interrelationships in nature: "Your mind would explode . . ." I'm thinking of Peacock on the wooded south slope, gathering fistfuls of orange chanterelles late in the autumn, not with any special electronic glove but barehanded.

"Educators are expected to use VR to create virtual textbooks in which students inhabit computerized recreations of the things they are studying," writes Potts. "Some visionaries even predict the advent of highly realistic computerized sex."

A brave and terrifying new world coming, and perhaps the trick is not to run from it, but to adapt to it. But if a place like the San Juans is important now, how much more important will it become in the future?

"It already is raising questions about whether virtual reality can be abused, like a drug, to create alternative psychedelic environments in which users could disappear from the real world," Potts explains.

"The problem . . . is that it's a door to another world," says NASA's virtual-reality expert, Michael McGreevy. "People might walk through it and choose not to come back."

What is our duty, for having been blessed with life for this brief moment? Is there a duty? I'm convinced there is. Taking care of one another, as well as of the land, is surely implicit in this partnership with life: with breathing, seeing, smelling, hearing, listening, speaking, thinking.

But there are wounds in this century that cannot be ignored, that must be doctored.

"Even in its more conservative forms, virtual reality can produce a disorienting and nauseating feeling known as 'simulator sickness,'" writes Potts, "in which the eyes tell the brain the body is moving, based on images from the VR system — contrary to the signals from the inner ear, which thinks the body is stationary."

Big George sits in the back seat and stares straight ahead into the night as the road rolls beneath us. We're going home, all of us, to different places, diverse places.

The San Juans and grizzly country: a land where nature still operates as a whole, as a system unfragmented by man, going at its own pace.

Thomas Furness, one of VR's leading researchers, says it bluntly: "We can change gravity, we can change the speed of light, we can change the speed of sound. We are all-powerful."

I do not believe it. We can change only the perception of these things. Nature, and the original system that created us, must always remain somehow with us, the bedrock of our movements and actions.

What is our duty? To live a life.

Sometimes I think the blue sparks I see are sparks of rage against what is being done to the earth. Other times I think they are sparks of joy, signs that I can barely stand to be alive, I love it so much.

Later in the night, George is talking about his father, who works for the government, teaching survival skills: mountaineering, fire building, water finding, meat hunting, gathering edible plants; what to do if your plane crashes and burns. Like his son, the elder Fischer has lived a life. He was a Russian prisoner of war during his sixteenth and seventeenth years. He's been around the world a number of times. "A lot of his friends have died up on the mountain," George says, speaking of "the mountain" as anyplace where there is danger, where there is hard reality. Anyplace where judgment and thought are necessary to survival.

George talks about preparation, about not cutting corners, about paying attention, about doing things correctly. I recall how anytime there is a thunderstorm, George never walks the ridges. "A bit of it rubbed off on me," George says, talking about his father's teachings. "It's that stupid little shit — you get away from it for a while" — going off without matches or rain gear or a compass or a map — "and then one day it catches your ass. *Pow!* You're dead, you're gone. You're not here anymore."

Scott rolls his window down to smell the sage in the high desert. A summer storm came through earlier in the night.

"Cottonwoods," George is saying, speaking almost to himself, perhaps remembering his father's advice. "Don't ever have any cottonwoods around your house. They're real pretty but their limbs are always falling off. They'll go right through your roof."

*

At three A.M. we stop for gas at a lonely, brightly lit convenience store on the outskirts of Price, Utah. The smell of rain and sage surrounds us. Inside is a fresh pot of coffee. We stagger around in the glare of fluorescent lights, purchasing junk the way a bear might browse berries from a bush, swallowing leaves, stems, berries and all.

Back in the car, the engine won't turn over, won't even click, though the battery's still strong. The lights and the tape player still work. A check under the hood reveals nothing.

"Maybe it needs to rest," Scott says.

We go back inside and sit down at one of the colorful butt-molded plastic-seat booths and drink a couple of cups of coffee. We wait about half an hour. Nobody else comes into the store during this time, and only one or two semis drive past.

We go back out and get in the car. I say, "Wait," before trying to start it. I'm at the wheel. Machines, with good reason, don't like me. "Everyone put on your seat belts. We've got to act totally and positively like this is going to work."

We all fasten our seat belts, like schoolchildren in a carpool.

I turn the key. The car trembles, shudders, then kicks into a purr and a hum, almost like a living thing.

*

"What about that young kid with the red hair?" Scott asks. "What was his name? Wasn't *he* a trip?"

"Dan," says George. "Yeah, Dan. Great. You could just feel that intensity."

"Like some kind of animal," Scott says.

"You could see it in his eyes," George says. "All that energy and life. He was like a marten. It's great to know he's out there, on our side. You could really see something behind his eyes."

*

Salt Lake glows orange, a corona of sunrise ahead of us. The Wasatch Mountains are blue and luminous, awakening the city below them. A city of a million people, I think, though the sunrise over it looks not all that different from the sunset I saw not twelve hours ago, in southern Colorado.

We're taking a back road that George knows about, cutting over from Durango to Dove Creek, driving through farmland in that slow, afterworldly orange twilight of long summer. Despite being in the middle of nowhere, there's fresh raw backhoe work running the length of the empty dirt road, with a house only every few miles and meadowlarks springing up from the fields as we pass.

"Fiber optics," George explains. "They're laying cable all across the country, like veins. Pretty cool stuff. You can link up anything, anywhere. It's almost like creating an organism, a kind of awareness or intelligence. So much information is going to end up being shared. It's almost like the old science-fiction movies — the system taking on a life of its own."

"Going to hook up every small town in America to it," Scott says. "Going to access everything and everyone."

On we cruise through the desolate, lovely country, alongside the open, waiting trenches. From the back seat, the smell of Mexican food from the restaurant in Durango fills the car.

✶

The creek at whose stony bottom you gazed as a child, and listened to, as the clear water rushed past — how far away is that? Is it still less than a lifetime away?

Can you still get there? Will you dare try?

Only a handful of grizzlies still exist, in Colorado and in a few other wild places in the West. They move around as a band, and at other times alone. Hounded by history, they often come out only at night. They skirt the high lakes above the tree line, their fur rippling, their muscles rolling. Heaven is in their teeth.

We must learn to love them. We have forgotten how to love them. Like all of us, they will not be here forever.

Bears? What bears?

In the late summer of 1995 a young man was walking high in the mountains, not far from where a Round River group had been active, when a large blond bear with stiff, dark-colored legs and long white claws hurtled over a ridge not forty feet away.

The man dropped to the ground and curled up in the fetal position, making a point of not looking at the bear. The bear ran toward the man and paced around him, circling, slamming its paws against the earth next to the man's body, which lay perfectly still. All the man could see was grizzly claws — over four inches long. The bear put its face up to the young man's face and blew slobber and hot breath on him; the man breathed it in but did not panic, did not get up and run.

Five times the bear stomped around the man, as if weaving a spell, all the while thumping the ground with its huge paws, making the ground tremble, snorting and roaring. Then the bear turned and walked off, having had his say. The young man, when he was sure the animal was gone, got up and ran all the way back to his camp.

They are still out there. We have not yet lost them. We stand only at the edge.